AMERICA IN THE 21ST CENTURY: POLITICAL AND ECONOMIC ISSUES

U.S. INNOVATION PROGRAMS

AMERICA IN THE 21ST CENTURY: POLITICAL AND ECONOMIC ISSUES

Additional books in this series can be found on Nova's website under the Series tab.

Additional E-books in this series can be found on Nova's website under the E-books tab.

SCIENTIFIC REVOLUTIONS

Additional books in this series can be found on Nova's website under the Series tab.

Additional E-books in this series can be found on Nova's website under the E-books tab.

AMERICA IN THE 21ST CENTURY: POLITICAL AND ECONOMIC ISSUES

U.S. INNOVATION PROGRAMS

ALLISON E. BERGER
AND
MARK C. KOEPEL
EDITORS

Nova Science Publishers, Inc.
New York

For permission to use material from this book please contact us:
Telephone 631-231-7269; Fax 631-231-8175
Web Site: http://www.novapublishers.com

LIBRARY OF CONGRESS CATALOGING-IN-PUBLICATION DATA

U.S. innovation programs / editors, Allison E. Berger and Mark C. Koepel.
p. cm. -- (America in the 21st century : political and economic
issues) (Scientific revolutions)
Includes bibliographical references and index.
ISBN 978-1-61122-809-0 (hardcover : alk. paper)
1. Technological innovations--Economic aspects--United States. 2.
Technological innovations--Government policy--United States. 3. Technology
and state--United States. I. Berger, Allison E. II. Koepel, Mark C. III.
Title: United States innovation programs.
HC110.T4U18 2011
338'.0640973--dc22
2010044746

Published by Nova Science Publishers, Inc. † New York

CONTENTS

PREFACE

In the United States, Congress authorizes most of today's federally-funded innovation inducement prizes beginning with the 108th Congress. This book explores federally-funded "innovation inducement prizes", the Technology Innovation Program (TIP) which was designed to support, promote and accelerate innovation in the U.S., the Small Business Innovation Research (SBIR) Program, issues in innovation and competition as well as tracking the state of innovation in the American economy.

Chapter 1- Since at least the 18th century, philanthropic organizations, industry, governments, and nongovernmental organizations throughout the world have offered many different kinds of prizes with a variety of objectives to reward accomplishments in science and technology. In the United States, Congress authorized most of today's federally-funded innovation inducement prizes beginning with the 108th Congress (2003). This analysis focuses on federally-funded "innovation inducement prizes," which are sponsored by federal organizations and designed to encourage scientists and engineers to pursue scientific and technical societal goals not yet reached.

Chapter 2- The Technology Innovation Program (TIP) at the National Institute of Standards and Technology (NIST) was established in 2007 to replace the Advanced Technology Program (ATP). This effort is designed "to support, promote, and accelerate innovation in the United States through high-risk, high-reward research in areas of critical national need," according to the authorizing legislation. Grants are provided to small and medium-sized firms for individual projects or joint ventures with other research organizations.

Chapter 3- In 1982, the Small Business Innovation Development Act (P.L. 97-2 19) established Small Business Innovation Research (SBIR) programs within the major federal research and development (R&D) agencies designed to increase participation of small innovative companies in federally funded R&D. Government agencies with R&D budgets of $100 million or more are required to set aside a portion of these funds to finance the SBIR activity. Through FY2007, over $22.3 billion in awards have been made for more than 100,016 projects. Extended several times, the program was scheduled to sunset on September 30, 2008. Although, to date, no specific legislation has reauthorized the program, the Small Business Administration determined that P.L. 110-235 temporarily extended the SBIR activity through March 20, 2009. Subsequently, P.L. 111- 10 provided an additional extension of the program through July 31, 2009; P.L. 111-43 extended it through September 30, 2009; and P.L. 111-66 extended the effort through October 31, 2009. The SBIR program was once again extended through April 30, 2010, by P.L. 111-89 and through September 30, 2010, by P.L. 111-214. On July 8, 2009, H.R. 2965, a bill to reauthorize and alter the SBIR initiative, passed the House; the Senate substituted the language of S. 1233 (amended) and passed its version of H.R. 2965 on July 13, 2009. For further information on SBIR reauthorization activity see CRS Report RS22865, *The Small Business Innovation Research (SBIR) Program: Reauthorization Efforts*, by Wendy H. Schacht.

Chapter 4- The Small Business Innovation Development Act of 1982, P.L. 97-219, created Small Business Innovation Research (SBIR) programs within the major federal research and development (R&D) agencies. This effort was intended to increase participation of small innovative companies in federally funded R&D. Government agencies with extramural R&D budgets of $100 million or more are required to set aside a portion of these funds to support research and development in small firms through the SBIR program. The program was extended several times and was scheduled to terminate on September 30, 2008. Although, to date, no specific legislation reauthorized the program, the Small Business Administration determined that P.L. 110-235 temporarily extended the SBIR activity through March 20, 2009. P.L. 111-10 provided an additional extension of the program through July 31, 2009; P.L. 111-43 extended it through September 30, 2009; and P.L. 111-66 extended the effort through October 31, 2009. P.L. 111-89 once again extended the SBIR program through April 30, 2010, after which P.L. 111-214 extended the effort through September 30, 2010. H.R. 2965, a bill to reauthorize and alter the SBIR initiative, passed the House after which the Senate substituted the language of S. 1233 (amended) and approved its version of H.R. 2965.

Chapter 5- Under current intellectual property laws, industrial designs may potentially be protected through design patents, trade dress, and copyright. In addition, the Vessel Hull Design Protection Act established a specialized, or *sui generis*, intellectual property right for the protection of boat hull designs. Some experts argue that the present intellectual property regime does not adequately protect industrial designers.

Chapter 6- The Advisory Committee on Measuring Innovation in the 21st Century Economy was established by the Secretary of Commerce in September 2006 to recommend ways to improve the measurement of innovation in the economy. In its report, the Advisory Committee outlines its recommendations to the Secretary of Commerce for steps to be taken by the government, the business community, and government and private sector researchers to foster and improve the measurement of innovation in the economy.

In: U.S. Innovation Programs
Editors: Allison E. Berger et al.

ISBN: 978-1-61122-809-0

Chapter 1

FEDERALLY FUNDED INNOVATION INDUCEMENT PRIZES

Deborah D. Stine

SUMMARY

Since at least the 18th century, philanthropic organizations, industry, governments, and nongovernmental organizations throughout the world have offered many different kinds of prizes with a variety of objectives to reward accomplishments in science and technology. In the United States, Congress authorized most of today's federally-funded innovation inducement prizes beginning with the 108th Congress (2003). This analysis focuses on federally-funded "innovation inducement prizes," which are sponsored by federal organizations and designed to encourage scientists and engineers to pursue scientific and technical societal goals not yet reached.

The objectives of such prizes are generally to identify new or unorthodox ideas or approaches to particular challenges; demonstrate the feasibility or potential of particular technologies; promote development and diffusion of specific technologies; address intractable or neglected societal challenges; and educate the public about the excitement and usefulness of research and innovation. They differ from "recognition prizes" such as the National Medal of Science, National Medal of Technology, and the Nobel prizes, which reward past S&T accomplishments.

The scientific and technological goals for federally-funded innovation inducement prizes include the full spectrum of research, development, testing, demonstration, and deployment. They are an alternative to more traditional ways of achieving societal objectives with S&T such as grants, contracts, fees, patents, and human or physical infrastructure investments that some think are too costly, risk-averse, and bureaucratic. Some believe that prizes, if designed well, can enhance the ability of science and technology to solve societal problems, by reaching a wider community of problem solvers, encouraging risk-taking, and focusing the attention of policymakers, entrepreneurs, the public, and researchers on the goals of an innovation program. Concerns about prizes are that they may inhibit the exchange of information among researchers and innovators due to the very nature of competitions, be challenging to design and finance, and result in duplicative work which may not be the best use of limited intellectual and financial resources.

Prizes differ in their intentions, objectives, sources of funding, competition mechanisms, reward structure, and other variables. The prizes themselves may take the form of recognition and publicity, cash, marketing monopolies, or other means. When a cash award is provided, most range from $250,000 to $2 million, can go up as high as $10 million, and have exceeded $500 million when the winner provides a service such as a vaccine. Some experts view the non- compensation portion of prizes such as recognition and publicity, as important, and sometimes more important, than the potential financial reward.

Members of Congress interested in federally-funded innovation inducement prizes may wish to consider several policy options including creating new prizes, and modifying or increasing oversight of current prize programs. In the 111th Congress, policymakers may make decisions that influence whether or not current prize programs will be funded, and existing programs modified. Some policymakers have proposed new prizes on technologies such as self-powered farms, voting systems designed for persons with disabilities, energy technologies, nanotechnology, cybersecurity, and automotive energy efficiency.

National governments throughout the world have offered prizes to encourage innovation since at least the late 1700s. For example, Napoleon's government offered a 12,000 franc prize for technologies that would enhance the preservation of food to better feed advancing military troops. This lead to the process of preserving food in bottles, which shortly thereafter led to the process of canned foods, and then broad use by consumers.[1]

In the United States, Congress authorized most of today's federally-funded innovation inducement prizes beginning with the 108th Congress (2003). The purpose of this chapter is to gain a better understanding of these prizes to provide guidance for Members of Congress who are interested in creating new prizes, modifying current prize programs, or increasing oversight of current prizes.

This chapter discusses the status of current federally-funded innovation inducement prizes, addresses the different types of prizes, analyzes when prizes may be appropriate and effective, and summarizes assessments that have been made of their effectiveness. The report also provides the lessons that may be learned from completed competitions, and policy options for those Members of Congress interested in taking action regarding federally-funded innovation inducement prizes. The report concludes with an overview of 111th congressional activities regarding prizes.

This chapter does not discuss prizes funded by non-federal organizations nor does it discuss recognition prizes that reward past accomplishments other than to distinguish them from innovation inducement prizes (see discussion of this issue in the following section, "What Are the Different Kinds of Prizes?").

What Are the Different Kinds of Prizes?

Philanthropic organizations, industry, governments, and nongovernmental organizations offer many different kinds of prizes with a variety of objectives to reward accomplishments in science and technology (S&T).[2] Some prizes, such as the Nobel prizes and U.S. National Medal of Science and National Medal of Technology, reward past accomplishments and do not have a specific scientific or technological goal. These have been called "recognition prizes." Other prizes, called "innovation inducement prizes," are designed to attain scientific and technical goals not yet reached, often in response to perceived market failures.

Objectives of these prizes include both technological and non-technological goals:

- Identify new or unorthodox ideas or approaches to particular challenges;
- Demonstrate the feasibility or potential of particular technologies;
- Promote development and diffusion of specific technologies;

- Address intractable or neglected societal challenges; and
- Educate the public about the excitement and usefulness of research and innovation.[3]

This chapter focuses upon federally-funded "innovation inducement" prizes that have these goals.

The scientific and technological goals for prizes include the full spectrum of research, development, testing, demonstration, and deployment. They are an alternative to more traditional ways of achieving societal objectives with science and technology such as grants, contracts, fees, patents, and human or physical infrastructure investments that some think are too costly, risk- averse, and bureaucratic. Some believe that prizes, if designed well, can enhance the ability of science and technology to solve societal problems, by reaching a wider community of problem solvers, encouraging risk-taking, and focusing the attention of policymakers, entrepreneurs, the public, and researchers on the goals of an innovation program. Concerns about prizes are that they may inhibit the exchange of information among researchers and innovators due to the very nature of competitions, be challenging to design and finance, and result in duplicative work which may not be the best use of limited intellectual and financial resources.[4]

Prizes differ in their intentions, objectives, sources of funding, competition mechanisms, reward structures, and other variables. There is also a wide spectrum of participants in prize competitions from individual citizens with and without scientific or technical expertise, school districts, governments, universities and other nonprofit organizations, and small and large companies. The prizes themselves may take the form of recognition and publicity, cash, marketing monopolies, or other means.[5] Some experts view the non-compensation portion of prizes as important, and sometimes more important, than the potential financial reward. From a competitor standpoint, key considerations are the degree of flexibility in the competition rules, and the financial and nonfinancial risks and incentives.[6]

Table 1. Federally-Funded Innovation Inducement Prizes

Agency	Competition	Technological Target	Total Prize	Status
Department of Defense (DOD)	**DOD Wearable Power Prize[a]**	Long-endurance, lightweight power pack for warfighters in the field.	$1.75 million.	Prizes awarded. A new competition is being considered.

	DARPA Grand Challenges[b]	Autonomous operation of unmanned ground combat vehicles	$3.5 million.	Competitions held in2004, 2005, 2007. Awards given in 2005 and 2007. No future competitions are planned.
Department of Energy (DOE)	**DOE Grand Challenges**	Breakthrough achievements in research, development, and commercial application that have potential for application to performance of DOE's mission.	$1- 10 million.	The three DOE Grand Challenge competitions, the Freedom Prize, H-Prize, and L-prize, are described in the following rows.
	• Freedom Prize[c]	Reduce country's dependence on foreign oil.	$1.5 million.	Competition expected to begin in 2009.
	• Hydrogen Prize (H-Prize)[d]	Hydrogen storage, and advancements in technologies, components or systems related to hydrogen storage.	$1 million.	Competition expected to begin in 2009.
	• Bright Tomorrow Lighting Prize (L-Prize)[e]	Three competitions: Replacements for 60 watt (W) incandescent light and parabolic aluminized reflector (PAR) 38 Halogen lighting; and a 150 lumens/watt (lm/W) "2 1st Century Lamp."	$10 million for 60W incandescent lamp category; $5 million each for PAR 38 and 21st Century Lamp categories.	Ongoing 2009 competition for 60W and PAR 38 replacements. Future competition expected for 21st century lamp.
	Progressive Automotive X PRIZE[f]	Clean, production-capable and super fuel efficient vehicles that exceed 100 MPG equivalent fuel economy. (MPGe)	$10 million from private sponsors; DOE provided $3.5 million for education activities.	Over 100 teams have registered for competition scheduled for 2010.
DOE and Environmental Protection Agency (EPA)	American Le Mans Series (ALMS) Green Challenge Race[g]	Encourage manufacturers to develop and introduce green technologies.	No financial prize. EPA and DOE provide in-kind support.	Two winners in 2008. In 2009, competition renamed Michelin Green X Challenge.

Table 1 (Continued)

NASA	NASA Centennial Challenge[h]	Drive progress in aerospace technology of value to NASA's missions, and find the most innovative solutions to technical challenges.	$300,000 to $2 million.	Six ongoing competitions (described in rows below). Future competitions on other topics are planned.
	• Astronaut Glove Challenge[i]	Improve glove design to reduce effort needed to perform tasks in space and improve the durability of the glove.	$250,000.	One competition held and won. Second competition in 2009.
	• General Aviation Tech nology[j]	Demonstrate the performance of light aircraft that incorporate improvements to maximize fuel efficiency, reduce noise, and improve safety.	$300,000.	NASA awarded a total of $97,000 in prizes in 2008. Competition scheduled for 2011 announced in 2009.
	• Lunar Regolith Challenge[k]	Design and build robotic machines to excavate simulated lunar soil.	$750,000.	Competition held in 2008 with no winner. New competition scheduled for 2009.
	Northrop Grumman Lunar Lander Challenge[l]	Build and fly a rocket-powered vehicle to perform simulated Lunar flight.	$2 million.	Level One of the competition completed in 2008, and $350,000 in prize money awarded. Level Two competition in 2009.
	Power Beaming and Tether ("Space Elevator")[m]	Two competitions: Power Beaming - Wireless power transmission; Tether - Exceed current tether strength.	$2 million.	Competitions held in 2006-2008 with no winner. Competitions scheduled for 2009.
	Lunar Oxygen Production or MoonROx[n]	Generate breathable oxygen from simulated lunar soil.	$1 million	Competition held in 2008 with no winner. Competition
				scheduled for October 2009.

HHS	**BARDA Project Bioshield**[o]	Effective medical countermeasures (e.g., diagnostic tests, drugs, vaccines, and other treatments) against chemical, biological, radiological, and nuclear (CBRN) agents.	Contract that guarantees government will purchase results of research and development proposed.	Ongoing competition with annual awards of contracts beginning in 2005. Awards thus far have ranged from less than $1 million to almost $900 million.

Source: Congressional Research Service based on information cited for each competition.

a. For more information, see http://www.dod.mil/ddre/prize/topic.html. Personal communication, CRS with Karen Burrows, DOD Prize Manager, March 27, 2009.

b. For more information, see http:// www. darpa. mil/ grandchallenge04/ sponsor_ toolkit/congress_lang.pdf; DARPA, DARPA Grand Challenge 2005:Rules, October 8, 2004 at http://www.darpa.mil/grandchallenge05/ Rules_8oct04.pdf; *http://www.d arpa.mil/grandchallenge04;* http://www.darpa.mil/grandchallenge05/; *http://www.darp a.mil/grandchallenge/index.asp;* and Personal communication, CRS with John Jennings, DARPA, on March 26, 2009.

c. For more information, see http://www.freedomprize.org/prizes/history.php. Personal communication, CRS with Karen Hanson, Executive Director, Freedom Prize, March 27, 2009.

d. For more information, see http://www. hydrogen. Personal communication, CRS with Jerry Hinkle, Technical Director, H-Prize, Technology Transition Corporation, March 31, 2009.

e. For more information, see http://www.lightingprize.org/index.stm.

f. For more information, see http://www.progressiveautoxprize.org/;

g. For more information, see http://www.epa.gov/OTAQ/ld-hwy/420f08031.htm; and http://www.americanlemans.com/index_green.php

h. For more information, see http://centennialchallenges.nasa.gov/; NASA FY2009 and FY2010 Budget Requests.

i. For more information, see http://astronaut-glove.tripod.com/.

j. For more information, see http://cafefoundation.org/v2/pav_home.php.

k. For more information, see http://regolith.csewi.org/.

l. For more information, see http://space.xprize.org/lunar-lander-challenge.

m. For more information, see http://www.spaceward.org/elevator2010-pb and http://www.spaceflightamerica.org/.

n. For more information, see http://moonrox.csewi.org/.

o. For more information, see CRS Report RL3 3907, *Project BioShield: Appropriations, Acquisitions, and Policy Implementation Issues for Congress*, by Frank Gottron.

WHAT IS THE STATUS OF FEDERALLY-FUNDED INNOVATION INDUCEMENT PRIZES?

The following federal agencies have science and technology (S&T) programs that conduct prize competitions: the Department of Energy (DOE), the Department of Defense (DOD) including the Defense Advanced Research Projects Agency (DARPA), the Department of Health and Human Services' (HHS) Biomedical Advanced Research and Development Authority (BARDA), and the National Aeronautics and Space Administration (NASA). Each of these agencies have the statutory authority to offer prizes. **Table 1** provides an initial overview, and the text that follows provides more in-depth information.

Department of Defense (DOD) Wearable Power Prize

The DOD Wearable prize was authorized by the John Warner National Defense Authorization Act of 2007 (P.L. 110-36), which stated that

> The Secretary of Defense, acting through the Director of Defense Research and Engineering and the service acquisition executive for each military department, may carry out programs to award cash prizes in recognition of outstanding achievements in basic, advanced, and applied research, technology development, and prototype development that have the potential for application to the performance of the military missions of the Department of Defense.

In response to this general authorization, DOD decided its first competition would be development of a long-endurance, lightweight power pack for warfighters in the field.

Competition Goals

The prize competition sought to inspire the use of ground-breaking and inventive approaches to solve technical problems; reach non-traditional DOD performers by lowering the barriers for participation; inspire students, academia, private inventors, and industry alike to leverage resources and compete using innovative ideas and approaches.[7] The winner of the contest was the lightest weight system weighing 4 kg or less at the weigh-in and meeting the total energy requirement as demonstrated in the competitive demonstration (bench plus field tests). **Figure 1** provides an overview of the prize's timeline, and may be illustrative of a typical prize timetable.

Of the completed competitions, the DOD Wearable Power Prize (which was managed by DOD with contractor support as needed) appears to have been the most successful in reaching a specific technological target for the federal government as well as enhancing its network of those interested in the topic, both internally within the services, and externally among possible contractors. DOD officials are discussing the next steps to advance the technology, not only with the winners, but the other participants as well.

DOD Assessment of Program

DOD has assessed the benefits of the program for itself and to prize competitors, and found that the competition provided several benefits. It helped validate the status and appropriateness of DOD investments, identify new approaches, create a national awareness of the importance of wearable power, facilitated the Pentagon and military Services working together to identify a joint direction for this technology before and after the competition, and identified seven organizations and groups new to working with DOD.[8]

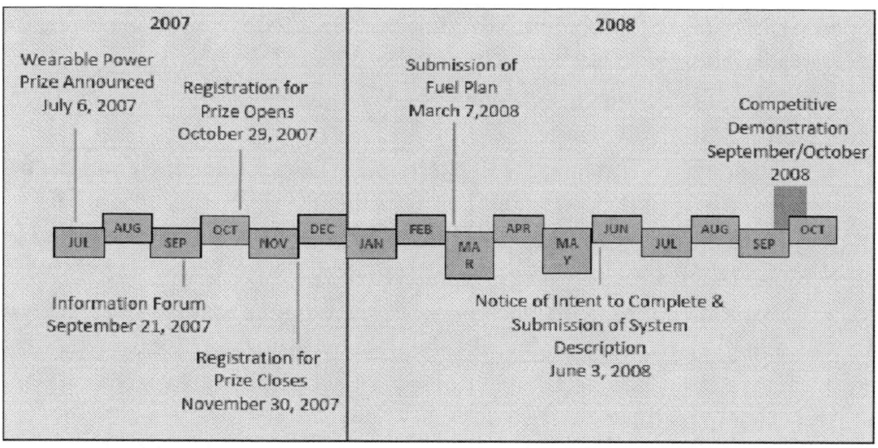

Source: DOD Wearable Power Prize Information Forum, powerpoint presentation, September 21, 2007 at http://www.dod.mil/ddre/prize/doc/WPP_IF_Brief9_21_07.pdf.

Figure 1. DOD Wearable Power Prize Timeline

DOD's assessment concluded that there were benefits to competitors, such as those participating in the competition were able to have access to DOD-paid and validated laboratory grade testing in close-to-operational conditions, and to DOD civilian and military professionals who provided direct feedback and real-time technical assessments. Competitors were also able to interact with other teams, which enhanced collaborative discussions and networking opportunities on topics of common interest. In addition, competitors received heightened national and international publicity through news reports and web activities.

Lessons for Future

DOD analyzed its competition to identify lessons learned for future competitions. According to DOD staff, among these lessons are—

- Choosing a topic or a competition goal that will attract the broadest public interest and ability to participate;
- Involving stakeholders (e.g., possible customers and competitors) from the beginning;
- Recognizing that setting competition metrics is critical;
- Deciding if topic addresses joint-service need (or not) and executing accordingly;
- Lowering competition entry and participation barriers to enable broadest involvement;
- Deciding if screening to determine whether concepts not deemed worthy of further consideration is prudent;
- Dedicating resources for media campaign and competitor communications (from program start to finish);
- Recognizing that a final public event requires significant resources; and
- Developing a post-competition plan that addresses expectations after the competition.[9]

This competition is concluded, but DOD is currently discussing at least one additional competition on a different technological challenge as part of its overall DOD prize program.[10]

Defense Advanced Research Projects Agency (DARPA) Grand Challenges

The DARPA[11] Grand Challenges were authorized in the Bob Stump National Defense Authorization Act for Fiscal Year 2003 (H.R. 4546, Sec. 2374b), which stated

> The Secretaries of the military departments and the heads of defense agencies may each carry out a program to award cash prizes in recognition of outstanding achievements that are designed to promote science, mathematics, engineering, or technology education in support of the missions of the U.S. Department of Defense.[12]

In response to the Floyd D. Spence National Defense Authorization Act for Fiscal Year 2001 (S. 2549, Sec. 217), which stated, "It shall be a goal of the Armed Forces to achieve the fielding of unmanned, remotely controlled technology such that by 2015, one-third of the operational ground combat vehicles of the Armed Forces are unmanned," DARPA decided to focus on autonomous robotic ground vehicles.[13]

Competition Goals

According to DARPA, the Grand Challenges sought to promote innovative technical approaches that would enable the autonomous operation of unmanned ground combat vehicles. These autonomous ground vehicles were to navigate from point to point in an intelligent manner to avoid or accommodate obstacles including nearby vehicles and other impediments. For the contest, DARPA held field tests of autonomous ground vehicles over realistic terrain and set specific performance goals for distance and speed. DOD planned to make three awards, first place for $2 million, second place for $1 million, and $500,000 for third place. [14]

The intent of the Grand Challenge program was to encourage participation by nontraditional partners so they might offer new, innovative ways of thinking that can lead to breakthroughs in various scientific or technological challenges. The cost of developing, fielding, and insuring entered vehicles was the sole responsibility of the individual teams. DARPA did not provide funding for the purpose of Grand Challenge entry or participation. Teams underwent a qualification process that included submission of the application, submission of an acceptable vehicle specification sheet and video demonstration, successful performance at the site visit, selection for the

National Qualification Event (NQE), submission of an appropriate technical paper and successful performance at the NQE. The NQE was the final qualification featuring a course that measures and tests vehicle capabilities where semifinalists vie for selection for the Grand Challenge Event.

Competitions

In 2004 and 2005, DARPA held Grand Challenges, and in 2007, DARPA hosted the Urban Challenge—an autonomous vehicle race through traffic. In 2004, participants were to develop vehicles that will navigate a course. No team entry successfully completed the designated route, and no award was made. [15]

In 2005, the DARPA Grand Challenge was similar to that in 2004. However, the test was in a different location that included 132 miles in desert terrain. Five teams completed the course, and first, second, and third place were awarded. [16]

In 2007, the DARPA Urban required teams to build an autonomous vehicle capable of driving in traffic, performing complex maneuvers such as merging, passing, parking and negotiating intersections. Eleven teams qualified and there were three winners.[17]

DARPA currently has no plans to hold an additional Grand Challenge event at this time. Should an additional challenge be held, it would likely focus on a different topic. [18]

DARPA Assessment of Program

According to DARPA, its Urban Challenge showed "breakthrough advances in autonomous vehicle capability and demonstrated for the first time autonomous vehicle operation in traffic," which is "being absorbed by the community, as expectations have been raised regarding autonomous vehicle capability and performance."[19] Teams that participated in the competition have begun identifying transition targets and partners. For example,

> Oshkosh Truck, which fielded Team Oshkosh Truck, has planned logistics demonstrations for the U.S. Army and U.S. Navy on vehicle platforms such as the Medium Tactical Vehicle Replacement, Palletized Load System, and Heavy Expanded Mobility Tactical Truck, and will demonstrate their vehicle for U.S. Army's Tank-Automotive Command Life Cycle Management Command.[20]

DARPA made the following overall assessment of its program:

The Urban Challenge program achieved its program goals and stimulated interest in the programs and projects of interest to the DoD Science and Technology (S&T) community. It was successful in attracting considerable joint investment by the participants and their sponsors, effectively leveraging Government investment in the program. The technical challenge was carefully defined and staged to bring coherence to the community and increase the chance for cross-fertilization among competing groups. The solicitation and qualification process was successful in attracting a large pool of strong teams with participation from the defense industry, automotive industry, academia, as well as a number of smaller organizations. This investment in expanding the community will continue to pay dividends as DoD benefits from a strengthened commercial sector autonomous vehicle technical community. The program has been successful in attracting many young people to work on S&T problems in areas affecting national security, and benefits are expected to accrue for many years as this group enters the work force.

The DARPA Grand Challenges in 2004 and 2005 made significant strides toward a day when autonomous robotic vehicles will perform hazardous tasks on the battlefield that today put America's fighting force in harm's way. In addition to saving lives, the technology will reduce stress on manpower requirements by requiring fewer support people. The DARPA Urban Challenge continued the acceleration of autonomous ground vehicle technology, making possible deployment on the battlefield within the timelines established by Congress.[21]

Department of Energy (DOE) Grand Challenges

The DOE Grand Challenges were authorized by the Energy Policy Act of 2005 (P.L. 109-58, Title X, Sec. 1008; EPACT 2005), in a section entitled "Prizes for Achievement in Grand Challenges of Science and Technology." This act states that "The Secretary may carry out a program to award cash prizes in recognition of breakthrough achievements in research, development, demonstration, and commercial application that have the potential for application to the performance of the mission of the Department." The Freedom Prize was created in the same act. The Energy Independence and Security Act (EISA) amended EPACT 2005 to create two additional prizes, the Hydrogen Prize (H-Prize) and the Lighting Prize (L-Prize). These prizes are scheduled to begin their activities in 2009.

Freedom Prize

The purpose of the Freedom Prize, authorized in EPACT 2005, is to encourage and recognize the development and deployment of processes and technologies that will improve America's national security, economic prosperity, and health by reducing the country's dependence on foreign oil.[22] The prize is to reward innovative deployment of existing technologies in five broad categories which include industry, military, schools, government and community. The first Freedom Prize competition, focused on school districts, is scheduled to begin in 2009. The Freedom Prize Foundation plans to give several awards, with total of $1.5 million in prizes.[23]

Hydrogen Prize (H-Prize)

The purpose of the H-Prize, authorized in EISA 2007 (Sec. 654), is to

> competitively award cash prizes in conformity with this subsection to advance the research, development, demonstration, and commercial application of hydrogen energy technologies.... The Secretary shall establish prizes under this subsection for—
>
> (i) advancements in technologies, components, or systems related to—
>
> (I) hydrogen production;
>
> (II) hydrogen storage;
>
> (III) hydrogen distribution; and
>
> (IV) hydrogen utilization;
>
> (ii) prototypes of hydrogen-powered vehicles or other hydrogen-based products that best meet or exceed objective performance criteria, such as completion of a race over a certain distance or terrain or generation of energy at certain levels of efficiency; and
>
> (iii) transformational changes in technologies for the distribution or production of hydrogen that meet or exceed far-reaching objective criteria, which shall include minimal carbon emissions and which may include cost criteria designed to facilitate the eventual market success of a winning technology.[24]

The 2009/2010 competition is to focus on storage materials for hydrogen in mobile systems for light-duty vehicles with a $1 million prize to the winner. Private contributions are expected to augment prize funds. The tentative schedule is to give notice of the competition in the *Federal Register* in Summer 2009, take entries by Fall 2009, and have a judging panel test and evaluate in Summer 2010. The $1 million prize is expected to be awarded in

Fall 2010. The Hydrogen Education Foundation administers the prize on behalf of DOE.

Bright Tomorrow Lighting Prize (L-Prize)

The goal of the L-prize, authorized in EISA 2007, is to spur development of ultra-efficient solid- state lighting products to replace the most widely used lighting, the 60-watt incandescent lamp and the PAR 38 halogen lamp, and to develop a "21st Century Lamp" that delivers more than 150 lumens/watt (lm/W). On June 24, 2009, DOE announced the beginning of the competition in a Federal Register notice.[25] Awards include cash prizes, subject to the availability of appropriated funds, of $10 million for the first successful product in the 60-watt incandescent lamp category and $5 million for the first successful product in each of the PAR 38 and 21st Century Lamp categories. In addition, there are opportunities for federal purchasing agreements, utility programs, and other incentives.

In this competition, companies do not enter the competition officially until they have 2,000 samples of the product ready for laboratory testing. The results of the testing will then be judged by a technical review committee, whose members may include utilities, lighting designers, and light-emitting diode (LED) technology experts, to determine the winning entry.[26] As of June 2009, it is expected that 5-10 companies are developing products for submission to the competition.[27] If no submissions are received by June 2010, then DOE has the option of either closing the competition or revising the standards under which it operates.

A unique feature of the L-prize is that there are a number of partners. Partners are organizations such as utilities and energy efficiency groups that have agreed through a memorandum of understanding to aid in marketing winning technologies. As of June 2009, there are approximately 23 partners who are active in 29 states.[28]

Progressive Automotive X PRIZE

The goal of the Progressive Automotive X PRIZE, managed by the X PRIZE Foundation with financial sponsorship by the Progressive Casualty Insurance Company, is "to inspire a new generation of viable, super-efficient vehicles that help break our addiction to oil and stem the effects of climate change."[29]

As noted earlier, when the federal government offers a prize, it frequently does so with the help of another organization who administers the prize either for a fee or at no cost through the provision of in-kind support. In this situation, another organization offers the prize and it is the federal government who provides support. An alternative approach is for the federal government to support a competition managed by a private or philanthropic organization, rather than administering the prize and acting as the primary sponsor itself.

A purse of $10 million in prizes would be awarded to the team(s) that develop a clean, production-capable and super fuel efficient vehicles that exceeds 100 MPG equivalent fuel economy (MPGe). This competition is focused on development of cars that would be made available for purchase, rather than concept cars. No technology is specified, but plans are to provide clear technical boundaries (i.e., for fuel-efficiency, emissions, safety, manufacturability, performance, capacity, etc.). In addition, the competition hopes to attract both existing automobile manufacturers and newcomers, and a balanced array of private investment, donors, sponsors, and partners to help competitors succeed (e.g., manufacturing assistance, testing resources, etc.). Beyond technological innovation, the prize sponsors plan to publicize the results, provide a cash award, and educate the public on key issues.

Over 100 teams representing 136 vehicles with 14 different fuel sources have passed the first judging stage allowing them to participate in the competition. The next stage is the judging of the vehicle designs to assess the vehicle's features, production capability, safety, and business plans. In the final stage, teams will compete in a long distance stage competition to assess vehicle performance and determine if their vehicles can exceed 100 MPGe. Both the competition events and the announcement of winners, which will take place after the event results have been analyzed, are expected in 2010.

DOE's Office of Energy Efficiency and Renewable Energy provided a $3.5 million grant to the Progressive Automotive X PRIZE for educational activities related to the competition. This includes a website, Fuel Our Future, developed with the X PRIZE Foundation and Discovery Education, that serves as an "interactive online portal offering stimulating science, technology, engineering and math (STEM) lessons and resources for students, teachers and families as the unique and engaging Progressive Automotive X PRIZE competition unfolds."[30] Additional plans for the funds include a national high school student contest, and educational events in host cities of the Progressive Automotive X PRIZE race series.[31]

American Le Mans Series (ALMS) Green Challenge Race

The Environmental Protection Agency's (EPA) Office of Transportation and Air Quality, DOE's Argonne National Laboratory, and the Society of Automotive Engineers (SAE) proposed a competition that incorporated "green" principles[32] into auto racing to encourage manufacturers to develop and introduce "green" technologies.[33] As with the Progressive Automotive X PRIZE discussed in the previous section, a non-federal organization sponsors the actual competition while federal agencies and others co-sponsors in-kind support. The Green Challenge Award sponsored by the American Le Mans Series (ALMS) with EPA, DOE, and the SAE as cosponsors, is for the fastest car with the smallest environmental footprint.[34] It is described further below.

The ALMS Green Challenge provides incentives for improved efficiency, use of renewable fuels, and reduced greenhouse gas emissions; allows any technology or fuel; and uses life-cycle analyses to assess both the on-track impacts and the upstream environmental and energy impacts of the fuel. These technologies were tested as part of the 1,000-mile Petit Le Mans race in the 2008 racing season.

Once the competition was in place, EPA, DOE, and SAE created the rules, regulations, and technical specifications for the competition as part of a "Green Racing Work Group." The government provided staff support for these activities. The ALMS funded the competition. Auto companies participating in the competition funded the research, development, and deployment of the cars in the competition. Thirty-seven cars competed for a trophy. There was no financial prize. Two winners were announced for prototype and grand touring (GT)[35] classes. According to one analyst, the competition accurately foreshadowed the ability of diesel injection technology to reduce emissions through the use of particle filters while maintaining high performance.[36]

The competition, renamed the Michelin Green X Challenge competition and sponsored by ALMS and the Michelin corporation, is taking place again in 2009.[37] The competitions began in the Spring of 2009, and will take place at each ALMS race in the 2009 racing season.[38] At the end of the season, EPA, DOE, and SAE International will present Green Challenge awards to the vehicle manufacturers in each class with the highest scores during the entire racing season.[39]

National Aeronautics and Space Administration (NASA) Centennial Challenges

The NASA Centennial Challenges were authorized in Section 104 of the National Aeronautics and Space Administration Authorization Act of 2005 (P.L. 109-155) which stated that

> The Administration may carry out a program to competitively award cash prizes to stimulate innovation in basic and applied research, technology development, and prototype demonstration that have the potential for application to the performance of the space and aeronautical activities of the Administration.

According to NASA, the Centennial Challenges seek to drive progress in aerospace technology of value to NASA's missions; encourage the participation of independent teams, individual inventors, student groups and private companies of all sizes in aerospace research and development; and find the most innovative solutions to technical challenges through competition and cooperation.[40] Individual challenges are either "first-to-demonstrate" competitions, or "repeatable contests" with prizes that range from $300,000 to $2 million. Each challenge is a public/private partnership with co-sponsor organizations that contribute cash towards the prize purse and allied organizations that provide in-kind services to enhance the competition. NASA's current challenges are described below.

Astronaut Glove Challenge

The goal of the astronaut glove challenge, managed by Spaceflight America, is to improve glove design to reduce effort needed to perform tasks in space and improve the durability of the glove. The 2007 challenge consisted of two competitions. One for a $200,000 prize, won by an unemployed aerospace engineer, reached its technological target of meeting, or exceeding, the specifications of NASA's current Phase VI glove. The winner subsequently started his own company to produce spacesuit gloves, and has a contract to provide gloves to another company that is producing spacesuits for the emerging private suborbital spaceflight industry. [41] The other $50,000 prize for a mechanical counter pressure glove went unclaimed.[42] The 2009 Astronaut Glove Challenge "is designed to promote the development of glove joint technology, resulting in a highly dexterous and flexible glove that can be used by astronauts over long periods of time for space or planetary surface excursions."[43]

General Aviation Technology

The general aviation technology competition, managed by the CAFE Foundation, involves a number of competitions with the goal of reducing the impact of aircraft on the environment, including demonstrating the performance of light aircraft that maximize fuel efficiency, reduce noise and improve safety. [44]

In 2007, NASA awarded $250,000 in prizes for personal air vehicles (PAV) that had the best performance as measured by a number of criteria including shortest runway, lowest noise, highest top speed, best handling qualities, and highest fuel efficiency, with $100,000 for the best overall performance. The 2008 General Aviation Technology Challenge included the Community Noise Prize ($150,000), Green Prize ($50,000), Aviation Safety Prize ($50,000), CAFE 400 Prize ($25,000)(a 400 mile cross-country air race), and Quietest Light-Sport Aircraft (LSA) Prize.[45] NASA awarded a total of $97,000 in prizes during this competition, with all but the Green Prize competitors receiving some level of financial award.[46] In 2009, draft rules were announced for the 2011 CAFE Aviation Green Prize with a proposed maximum purse of $1.7 million dollars.[47]

Lunar Regolith Excavation Challenge

The goal of the lunar regolith[48] excavation challenge, managed by the California Space Authority, is to design and build robotic machines to excavate simulated lunar soil. The winning team will receive a prize of $750,000. Twenty-five teams have registered for the competition. Sixteen teams competed in 2008, but no team was able to win the prize. The next competition is scheduled to take place on August 15, 2009.[49]

Northrop Grumman Lunar Lander Challenge

Competitors in the Northrop Grumman Lunar Lander Challenge, managed by the X PRIZE Foundation, with a total of $2.0 million in prize money available, must build and fly a rocket-powered vehicle under conditions that simulate the flight of a vehicle on the Moon.[50] From 2006- 2008, three competitions were held on fixed dates in fixed locations. In 2008, level one of the competition was won with the winner receiving $350,000. In 2009, competitors will seek to win the remaining $1.65 million in level two of the competition at a date and location of their choosing between July 20, 2009, to

October 31, 2009. Nine teams have registered for level two of the competition including five veteran teams, and four new teams. Winners are determined by the X PRIZE Foundation.

Power Beaming and Tether

The Power Beaming and Tether prize, managed by the Spaceward Foundation, includes two competitions, which together are called the Space Elevator Games.[51] In the power beaming competition, the goal is for teams to provide a practical demonstration of wireless power transmission by building mechanical devices that propel themselves up a vertical cable while the power supply remains on the ground. Competitions were held in 2006, 2007, and 2008 with no winners of the prize purse of $2 million awarded. A 2009 competition is planned for July. In the August 2009 tether competition, an additional $2 million prize purse will be given to the team that develops the material for a tether that can exceed the strength of the best available commercial tether by 50% with no increase in mass.

Lunar Oxygen Production or MoonROx

The MoonROx Challenge, managed by the California Space Education and Workforce Institute, has a technological goal of developing a process to extract breathable oxygen from lunar regolith.[52] A competition was held in 2008 with no winners of the $1 million purse. The next competition is scheduled for October 2009.

NASA Assessment of Program

The targeted outcomes for the NASA Centennial Challenges are to drive progress in aerospace technology of value to NASA's missions; encourage the participation of independent teams, individual inventors, student groups, and private companies of all sizes in aerospace research and development; and find the most innovative solutions to technical challenges through competition and cooperation. In its FY2009 budget request, NASA stated that the outcome of the program is to be evaluated based on its ability to "demonstrate benefits of prize competitions by awarding at least one prize and communicating the resulting technology advancements."[53]

According to NASA,

> Overall, the amount of team diversity (representing small and large businesses, high school and university students, and enthusiastic hobbyists and garage mechanics) and the variety of technologies implemented exceeded

> Agency expectations. As the prize purses increase, the amount of participation and level of technical maturity and ingenuity will also increase. In the past competitions where the prize purses were on the order of $300,000 each, it is estimated that the 10-15 participating teams represented an investment of $50,000 -$100,000 each. In the competition with a $2 million prize purse, teams invested on the order of $250,000 - $500,000 each.

In addition, NASA states that "Centennial Challenge competitions have spurred the creation of new businesses and products, including innovations in pressure suit gloves and reusable rocket engines."[54]

NASA makes the following assessment of the Centennial Challenge competitions:

> Prize programs encourage diverse participation and multiple solution paths. A measure of diversity is seen in the geographic distribution of participants (from Hawaii to Maine) that reaches far beyond the locales of the NASA Centers and major aerospace industries. The participating teams have included individual inventors, small startup companies, and university students and professors. An example of multiple solution paths was seen in the Regolith Excavation Challenge. NASA can typically afford one or two working prototypes but at this Challenge event, sixteen different working prototypes were demonstrated for the NASA technologists. All of these prototypes were developed at no cost to the government.
>
> The return on investment with prizes is high as NASA expends no funds unless the accomplishment is demonstrated. NASA provides only the prize money and the administration of the competitions is done at no cost to NASA by non-profit allied organizations. For the Lunar Lander Challenge, twelve private teams spent nearly 70,000 hours and the equivalent of $ 12 million trying to win $2 million in prize money. Prizes also focus public attention on NASA programs and generate interest in science and engineering. During the recent Lunar Lander Challenge, a live webcast had over 45,000 viewers and over 100,000 subsequent downloads. Prizes also create new businesses and new partners for NASA. The winner of the 2007 Astronaut Glove Challenge started a new business to manufacture pressure suit gloves. Armadillo Aerospace began a partnership with NASA related to the reusable rocket engine that they developed for the Lunar Lander Challenge, and they also sell the engine commercially.[55]

Future Competitions

NASA indicates that in selecting topics for future NASA Centennial prize competitions, it will consult widely within and outside of the federal government, and use selection criteria that include addressing common NASA and national technology needs; balancing the challenges across the fields of science, exploration, space operations, and aeronautics; and broadening the geographical distribution of competitor teams and host venues.[56] It is considering future challenges focused on revolutionary energy storage systems, solar and other renewable energy technologies, laser communications, demonstrating near-Earth object survey and deflection strategies, innovative approaches to improving the safety and efficiency of aviation systems, closed-loop life support and other resource recycling techniques, and low-cost access to space.

Biomedical Advanced Research and Development Authority (BARDA) Project BioShield

BARDA[57] manages Project BioShield,[58] which is to accelerate the research, development, purchase, and availability of effective medical countermeasures (e.g., diagnostic tests, drugs, vaccines, and other treatments) against chemical, biological, radiological, and nuclear (CBRN) agents. Some view Project BioShield as a prize competition, where the prize is the award of a contract that guarantees the government will purchase the results of the research and development proposed.

Project Bioshield was authorized in the Project BioShield Act of 2004, which has three main provisions. One of these creates a government-market guarantee by allowing the HHS Secretary to obligate funds to purchase CBRN countermeasures while they still have several more years of development. However, companies may receive payment when development is complete and the product is delivered. The Pandemic and All-Hazards Preparedness Act (PAHPA; P.L. 109-417) modified the Project BioShield Act to allow for milestone-based payments for up to half of the total award before countermeasure delivery. Awards thus far have ranged from less than $1 million to almost $900 million.

HHS noted that several awards have resulted in products being added to the Strategic National Stockpile (SNS). However, HHS also noted that one award was cancelled and other opportunities have gone unfilled.[59] In 2007, Project Bioshield used the original Project BioShield 10% advance payment provision as well as milestone payment authorities provided by PAHPA. In 2008, BARDA issued a Request for Proposals (RFP), but had to cancel it in FY2009 due to the immaturity and excessive risk associated with awarding contracts to the organizations that submitted proposals, but later issued a different RFP.[60]

What Policy Options Might Members of Congress Consider?

Members of Congress interested in federally-funded innovation inducement prizes may wish to create new prizes, or modify or increase oversight of current prize programs. In discussing possible options, policymakers may wish to receive comments from the current or potential administering agency, and stakeholders in that prize including participants in the competition, sponsors of those competitors, organizations that partner with the agencies in the administration of a prize, and the users of the competition results from both a technological and educational perspective.

Create New Prizes

Policymakers interested in exploring the possibility of additional prizes may be interested in the conclusions of several studies examining whether or not prizes are an appropriate policy mechanism to reach a particular societal goal relative to alternative research and development mechanisms; prize design, administration, and financing; and other possible considerations when developing prize legislation. Each of these issues is discussed further below.

Goals

As discussed earlier, prizes can have a number of goals. Although the primary focus may be achieving a scientific or technological goal including identifying new approaches to a challenge, some believe that the other aspects of competitions for a prize may be even more important. These subsidiary goals generally focus on the publicity surrounding prizes that may encourage the diffusion of specific technologies; bringing attention to intractable or neglected societal challenges; educating the public, particularly students, about the excitement and usefulness of research and innovation; and stimulating effort across the spectrum of research and innovations, including basic research, technology deployment and diffusion, and managerial/organizational innovation.[61]

Agencies also discussed the benefits of prizes for reaching goals such as expanding the network of academic and non-academic researchers willing to work with a federal agency, and enhancing communication among researchers who currently work with an agency and within an agency or between multiple agencies.

In considering legislation for prizes, policymakers may wish to take actions to identify broader goals such as these and encourage mechanisms for achieving them. For example, if policymakers are interested in using prizes as a mechanism for educating students about science and engineering, they may wish to state this goal specifically and encourage agencies to take actions that would enhance the ability of reaching this goal.

Appropriateness and Design

Policymakers interested in prizes may wish to consider the results of the following studies on innovation inducement prizes (both private and public sector), which focus on whether or not a prize is appropriate, and how a prize can be designed to best reach the sponsor's goal. These studies reached the following, sometimes contradictory, conclusions:

- *National* Academy of Engineering: Prizes, if designed well, can reach a wider community of problem-solvers than grants, and are useful when the desired output is not patentable. They are also useful when typical R&D funding mechanisms are too risk-averse, costly, or bureaucratic; or when there s an inadequate or nonexistent private market. Prizes can be tailored more precisely than other reward systems, though they may be challenging to design, determine awards, and finance. [62]
- *National Research Council*: Prizes can focus the attention of policymakers, entrepreneurs, the public, and researchers on the goals of an innovation program and reduce the administrative burden of grants and contracts. However, they may be less suited to attain certain societal goals than grants and contracts. For example, prizes may not be as suitable as grants and contracts in enhancing basic scientific and engineering understanding, and may put less emphasis on educating and training the next generation of researchers. In addition, prize competitions are likely, by their very nature, to inhibit

the exchange of information among researchers and innovators at least for the duration of the project. Participation by would-be innovators also may be inhibited due to insufficient funds to participate in the prize competition.[63]

- *Resources for the Future*: Appropriate design is critical to prize success based on theoretical and historical evidence, so interested parties from both the public and private sectors should carefully consider how their specific resources and goals relate to the different prize design elements. If prizes are put into place on an experimental basis, it is important to understand the actual effectiveness and efficiency of the prize program, and how their design influences the results they achieve. Examples of design elements include the institutional setting (e.g., public or private sector), technological target relative to the sponsor's overall goals, financial award (e.g., prize should not be so large that it leads to excessive research, but not so small that researchers do not pursue), and victory conditions (e.g., "first past the post," where the first competitor to reach pre-determined award criteria wins the prize, which emphasizes speed and an explicit target; or "best in class," where everyone competes on the same day, which provides more flexibility for maximizing achievement in a given timeframe).[64]
- *McKinsey & Company*: When conducted in an open, competitive, and media- friendly way, prizes can be a unique and powerful tool that identifies new levels of excellence, encourages specific innovations, changes wider perceptions, improves the performance of communities of problem-solvers, builds individual skills, and mobilizes new talent and capital. However, designing and delivering successful prizes is hard work. Prizes may be most useful when there is a clear goal, a relatively large number of potential solvers willing to absorb some of the risk, and a range of success criteria that includes a clearly-defined societal benefit that can be translated into prize objectives that are significant, motivational, actionable, results-focused, and time-bound.[65]
- *Brookings Institution:* Prizes are especially suitable when the goal can be defined in concrete terms, but the means of achieving the goal are too speculative to be reasonable for a traditional research program or procurement. Prizes also offer the potential for allowing government to establish a goal without being prescriptive as to how that goal should be met; and can stimulate philanthropic and

private sector investment that is greater than the cash value of the prize and attract teams with fresh ideas who might not otherwise do business with the federal government. Prizes, however, have significant limitations including the challenges faced by entrants in having or raising funds to compete, or the difficulty of clearly specifying in advance the victory conditions and quantifiable fundamental research outcomes. In addition, competitions lead to multiple research teams working on the same project, which may not be the best use of limited intellectual and financial resources.[66]

- *Centre for Economic Policy Research*: Innovation inducement prizes can be a powerful mechanism for encouraging competition, and prestigious non-financial prizes can be particularly effective at encouraging innovation, but may have high administrative costs.[67]

Administration

Administering a prize includes not only the scope of the prize as discussed above, but such issues as deadlines, prize administrators, prize financing, competitor selection, judging procedures, intellectual property,[68] liability,[69] and public relations.[70] The cost of administering a prize may exceed the financial reward given to winners of the prize.

In developing legislation, policymakers may wish to consider whether or not they want a federal agency to administer a prize on its own, to have a separate organization such as a nonprofit organization administer a prize on behalf of the federal agency, or to support the prize competition of another organization.

Some experts believe that a growing prize industry, familiar with prize administration, may provide enhanced outcomes as it professionalizes the prize process by bringing more formal knowledge and experience into the design and management process.[71] However, there are also some challenges in taking this approach as an organization may have less control over the quality of the judging. This approach may also make it more difficult for organizations to integrate prize activities with other supporting activities such as related science and technology activities as well as lectures, student activities, and other public events that may be managed by other parts of the sponsoring organization.[72]

Financing

The financing of a prize includes both the prize purse that may be provided to competitors, and the cost of administering the prizes. In some cases, the cost of administering a prize can exceed the cost of the prize awarded to the competitor. Note that in some cases there is no financial reward, so the only cost is the administration of the prize.

Prizes are financed through a variety of mechanisms. Some prizes are totally funded and staffed by the federal agency. Other prizes are funded by a federal agency, but staff support (sometimes for no or limited cost) is provided by outside, nonprofit organizations. In other cases, prizes are funded by a private organization, and the federal agency provides staff support.

Competitors, in general, receive no funding from either the federal agency or private organizations involved in the prize competition. They must seek financial sponsorship. In cases where a large company is competing for a prize, this may not be a major issue. In situations where universities, small companies, or nonprofit organizations are competing, the degree of financing available to them may determine whether or not they can compete, or the level at which they compete.

Although a prize may be authorized by Congress, unless there is a specific appropriation, the competition may not take place. A related issue is whether holding the prize competition is an option or a requirement. An authorized prize competition may not occur if an agency does not request specific funding for a prize, or Congress does not make a designated appropriation. In some cases, agencies set aside funding from existing programs or solicit it from programs throughout the agency. The higher the financial reward, the more challenging such actions are to take.

Difficult economic conditions may raise additional financial policy issues regarding the previously discussed in-kind or financial sponsorship for competitors. Such conditions may limit the availability of non-governmental partners willing to provide low or no cost management for the prizes. Similarly, these conditions may make it more difficult for competitors to find financial sponsors.

Another financial consideration for prizes is what occurs to funding set aside for a financial prize when no competitor meets the technological conditions necessary to win the prize within a specified timeframe. Does the prize funding move to a new competition for the following year? Or does a policy decision need to be made as to whether the competition should end, or should the technological conditions be revised so that competitors potential of winning a prize is enhanced? Some competitions provide an option to revise conditions and prize funding each year.

Legislation Considerations

In sum, policymakers may want to consider their responses to the following questions when developing legislation for a federally-funded innovation inducement prize:

- Should the legislation be general, providing federal agencies with an overview of the prize goals, or specific, detailing instructions to the agency regarding the prize competition? Such details may include the goals of the program, timeframe, award, participant eligibility, administration, contest rule determination, competition judges, intellectual property, liability, and program evaluation.
- How much flexibility should the agency have in determining the prize goals and in the administration of the prize?
- What should be the prize topic? Who should select it? (See **Box 1**.)
- What should be the goals of the program? What is the relative importance, for example, of technological advancement, education, and public awareness?
- Should there be a timeframe associated with the prize?
- Should there be a monetary award associated with the prize? If so, for how much? If not, is the publicity associated with winning the prize sufficient to encourage quality contestant participation? Should there be intermediary prizes?
- Who should be eligible to participate in the competition? For example, should employees of federal agencies or Federally Funded Research and Development Centers (FFRDCs) or Government-Owned, Contractor-Operated (GOCO) laboratory employees be allowed to compete? If so, can they use federal funds and facilities? Should foreign entities, such as non-U.S. citizens, corporations, or U.S. subsidiaries of foreign-owned corporations, be allowed to compete?
- Who should administer the program? For example, should a federal agency administer the program independently, do so in partnership with other federal agency or non-federal organizations, or act as a financial or non-financial partner in a competition administered by a non-federal organization?
- Should the competition be “first-past-the-post” or “best in class”?

- What happens to a financial prize if there is no winner?
- Who should determine the contest rules?
- Who should judge the competition? Should there be an appeal process?
- What are the criteria for the program to be considered successful?
- Should the program be evaluated? If so, by whom? When should that organization become involved?

Answering some of these questions may be challenging for policymakers without the guidance of the science and technology community and those experienced in the administration of prize competitions. In addition, the constantly changing nature of science and engineering due to new discoveries and innovation may also influence the need to provide sufficient flexibility in prize legislation.

Modify Current Prize Programs

Congress may decide that an existing program needs modification to meet national goals. The legislation authorizing prizes varies. Some legislation is specific in terms of how the prize is managed and how the winner is determined, while in other cases, the legislation is general and leaves such decisions up to the agency administering the prize. In addition, some prizes are authorized for only a specific period of time, and Members of Congress may wish to modify that timeframe.

Members of Congress may wish to modify the management and conditions under which a prize is given based on information they receive from stakeholders. Examples of possible stakeholders include organizations and individuals from universities, schools, science and engineering societies, trade associations, business, industry, venture capital groups, early-stage investors, philanthropic organizations, nongovernmental organizations, and foundations.

Box 1. Criteria for Selection of Prize Topics

A National Research Council committee identified the following criteria for selecting prize topics:

Criteria Related to Government Encouragement of Innovation

- The contest goal is widely judged to be worth pursuing and is in fact among the most important challenges facing the nation.
- If a prizewinning innovation is developed and put into practice, it will offer substantial practical benefit not only to its producers and users but also to the nation as a whole.
- Pursuit of the innovation should be perceived as a high-risk but high-reward activity.
- Without government intervention the market is unlikely to produce the innovation in a timely or effective manner, that is, the usual arguments from market failure for an affirmative federal involvement in the innovation process should apply to prize programs just as they do to other programs and policies intended to encourage innovation.

Criteria Related to the Use of the Inducement Prize Mechanism

- The prize goal should represent an ambitious effort, well beyond the current state of the art.
- It is expected that the contest objective can be achieved within a reasonable time frame for a prize program; on the order of 2 to 10 years.
- It will be possible to determine in a relatively objective manner whether a particular contestant's innovation has in fact achieved the contest objective.
- It will be feasible to define a plausible contest objective that is a suitable surrogate for the test of innovative success that is usually applied in the marketplace.
- The contest will encourage a wide range of types of contestants, including those not ordinarily active in the research grant and contract world, to participate.
- The goal is unlikely to be achieved, at least not in a reasonable time, using traditional grant and contract modes of encouragement alone.

Criteria Related to Broad Outreach and Engagement

- The goal should be reasonably meaningful to the general public and understandable by a wide range of potential contestants.
- The methods and tools required to compete effectively should be available to a reasonably large number and wide range of potential contestants.
- The process of competing for the prize should encourage formation of new social networks among individuals, firms, government laboratories, financial institutions, and others, who can contribute to future innovative activities.

Criteria Related to Political and Social Constraints

- Achieving contest objectives should not require use of classified information or technologies nor should it result in inadvertent creation of same.
- Pursuing the prize should not pose unreasonably large risks to contestants, [the federal government,] or the larger society.

Source: National Research Council, *Innovation Inducement Prizes at the National Science Foundation* (Washington, DC: National Academy Press, 2007).

Increase Oversight of Current Prizes

If policymakers are interested in increasing oversight of a federally-funded innovation inducement prize, they may wish to focus on whether or not the goals of the prize were achieved and its objectives clearly communicated, and how well the program was administered. There have been few external studies of prize effectiveness relative to alternatives, particularly federally- funded prizes. The prizes currently in place, therefore, provide an opportunity for policymakers to determine through their oversight mechanisms whether existing prizes should be discontinued or enhanced, and new prizes authorized. Possible prize evaluation criteria, by category, for evaluating the program include whether the contest—

Supply of Technology

- enhanced advanced innovation, or led to related innovations;
- identified new or unorthodox ideas or approaches to particular challenges;
- demonstrated the feasibility or potential of particular technologies;
- solved a challenging, well-defined problem requiring innovation;
- highlighted a range of best practices, ideas, or opportunities within a field;
- focused attention on, set standards in and/or influence perception of a particular field or issue;

Demand for Technology

- promoted development and diffusion of specific technologies;
- emulated market incentives, driving down costs through competition and exposing latent demand;

Scientific and Technical Community

- educated and changed behavior of participants through the prize process;
- celebrated and strengthened a particular community;
- registered large numbers of contestants, and those from a more diverse group than the traditional constituency for that agency;

Public Awareness

- educated the public about the excitement and usefulness of research and innovation.
- enhanced public awareness of and interest in the issue or federal agency

Prize Administration

- addressed intractable or neglected societal challenges;

- attracted private sources of funds to support the research activities of contestants, augment the prize purses, or provide in-kind services;
- functioned well with appropriate prize rules and processes.[73]

ACTIVITIES IN THE 111TH CONGRESS

This section describes the activities in the 111th Congress regarding federally-funded innovation inducement prizes.[74]

In the Omnibus Appropriations Act, 2009 (P.L. 111-8), Congress indicated that funds for announced, authorized prizes at NASA shall remain available, without fiscal year limitation, until the prize is claimed or the offer is withdrawn. During the FY20 10 appropriations process, NASA, requested $4.0 million for its Centennial Challenges program. No funding was provided in FY2009. In H.Rept. 111-149, the House recommended not providing this funding due to "affordability considerations."

Several Members of Congress have also introduced legislation in the 111th Congress that would establish prizes on several science and engineering topics.[75] Each of these is described below.

H.R. 41 would provide for federal research, development, demonstration, and commercial application activities to enable the development of farms that are net producers of both food and energy, and for other purposes. It would direct the Secretary of Energy to enter into an arrangement with the National Academy of Sciences to evaluate the feasibility of a prize and best practices award programs as tools to promote self-powered farms and, if feasible, make recommendations for carrying out such programs.

The Voting Integrity and Verification Act of 2009 (S. 48) would require the Director of the National Institute of Standards and Technology (NIST) to establish a program to award cash prizes competitively to eligible persons that advance the research, development, demonstration, and application of voting systems specifically designed to enhance accessibility and provide independence for persons with disabilities during the voting process.

The New Manhattan Project for Energy Independence Act (H.R. 513) would award a prize for the development and manufacturing of energy technologies that meet a number of economic, technical, and societal criteria including a plug-in hybrid vehicle, alternative fuel vehicle, electric vehicle, hydrogen fuel cell vehicle, or other alternative technology vehicle; an energy efficient residential or commercial building; a large scale solar thermal power plant or solar photovoltaic power plant; biofuels; carbon capture and storage system for a large scale coal-burning power plant; validated process for remediation of radioactive waste; or sustainable nuclear fusion reaction. The DOE would administer a prize program with monetary awards to advance the research, development, demonstration, and commercial application necessary to achieve these goals with a prize for each. Other goals include broad participation by researchers, large and small businesses, institutions of higher education, and any other qualified applicants, including veterans. In developing the prize, DOE is to consult with other federal agencies, including NSF; and may consult with other experts such as private organizations, including professional societies, industry associations, and the National Academy of Sciences and the National Academy of Engineering.

The Nanotechnology Innovation and Prize Competition Act of 2009 (S. 596) directs the Secretary of Commerce, through the NIST Director to establish a program to award prizes for achievement in the applications of nanotechnology for: (1) improvement of the environment, consistent with Twelve Principles of Green Chemistry of the Environmental Protection Agency (EPA); (2) development of alternative energy that has the potential to lessen the dependence of the United States on fossil fuels; (3) improvement of human health, consistent with regulations promulgated by the Food and Drug Administration (FDA); and (4) development of consumer products. The bill would appoint a board to oversee the program and require it to submit an annual report to Congress. It provides the board with the option of contracting with a private organization to administer the prize. That board could either make a financial award to the first competitor to meet criteria established by the board, or recognize individuals for superlative achievement in one or more of these nanotechnology applications and recommend to the Secretary of Commerce that the President award the National Technology and Innovation Medal (which would not have a financial award).

The Cybersecurity Act of 2009 (S. 773) would require the NIST Director to establish cybersecurity competitions and challenges with cash prizes to attract, identify, evaluate, and recruit talented individuals for the Federal information technology workforce, and stimulate innovation in basic and applied cybersecurity research, technology development, and prototype demonstration that have the potential for application to the federal information technology activities of the federal government. The program would target its competitions and challenges to different groups including high school students, undergraduate students, graduate students, and academic and research institutions. Competition topics are to be developed by the NIST

Director based on consultations with organizations both within and outside the federal government, and may establish advisory committees to provide additional guidance. To carry out this program, $15 million would be authorized to be appropriated to NIST for each of FY20 10 through FY20 14.

The New Options Petroleum Energy Conservation Act of 2009 (H.R. 1794) would require that the Secretary of Energy establish a program to award a $1 billion prize to the first automobile manufacturer incorporated in the United States to manufacture and sell in the United States 60,000 midsized sedan automobiles that operate on gasoline and can travel 100 miles per gallon.

End Notes

[1] Dale Blumenthal, "The Canning Process: Old Preservation Technique Goes Modern," *Food and Drug Administration Consumer Magazine*, September 1, 1990, at *http://www.fda .gov/bbs/topics/CONSUMER/CON00043.html.*

[2] For lists of some existing prizes, see Knowledge Ecology International, *Selected Innovation Prizes and Reward Programs*, KEI Research Note 2008:1 at http://www.keionline.org/misc-docs/research_notes/kei_rn_2008_1.pdf; and McKinsey & Company, *And the Winner is ... Capturing the Promise of Philanthropic Prizes*, 2009 at *http://www.mckinsey.com/ clientservice/socialsector/And_the_winner_is.pdf.*

[3] National Academy of Engineering, *Concerning Federally Sponsored Inducement Prizes in Engineering and Science* (Washington, DC: National Academy Press, 1999).

[4] National Academy of Engineering, *Concerning Federally Sponsored Inducement Prizes in Engineering and Science* (Washington, DC: National Academy Press, 1999) at http://www.nap.edu/catalog.php?record_id=9724; National Research Council, *Innovation Inducement Prizes at the National Science Foundation* (Washington, DC: National Academy Press, 2007); Richard G. Newell and Nathan E. Wilson, *Technology Prizes for Climate Change Mitigation*, RFF DP 05-33, Resources for the Future, June 2005 at http://www.rff.org/documents/RFF-DP-05-33.pdf; McKinsey & Company, *And the Winner is ... Capturing the Promise of Philanthropic Prizes*, 2009, at http://www.mckinsey.com/ clientservice/socialsector/And_the_winner_is.pdf; Thomas Kalil, *Prizes for Technological Innovation*, The Brookings Institution, December 2006 at *http://www.brookings.edu/ views/papers/200612kalil.pdf;* Liam Brunt, Josh Lerner, and Tom Nicholas, *Inducement Prizes and Innovation,* CEPR Discussion Paper No. DP6917, July 2008 at http://ssrn.com/ abstract=1307507.

[5] Knowledge Ecology International, *Selected Innovation Prizes and Reward Programs*, KEI Research Note 2008:1 at *http://www.keionline.org/misc-docs/research_notes /kei_rn_2008_1.pdf.*

[6] Barry J. Nalebuff and Joseph E. Stiglitz, "Prizes and Incentives: Towards a General Theory of Compensation and Competition," *The Bell Journal of Economics* 14(1): 21-43, Spring 1983.

[7] For more information, see http://www.dod.mil/ddre/prize/topic.html.

[8] John W. Hopkins, Project Manger, Army Research Laboratory and Karen S. Burrows, Defense Research & Engineering Prize Manager, Wearable Power Prize Competition, "Wearable Power Prize Competition," powerpoint presentation, December 11, 2008.

[9] Ibid.

[10] Personal communication, CRS with Karen Burrows, DOD Prize Manager, March 27, 2009.

[11] DARPA is located at the Department of Defense. For more information, see CRS Report RL34497, *Advanced Research Projects Agency - Energy (ARPA-E): Background, Status, and Selected Issues for Congress*, by Deborah D. Stine.

[12] For more information, see http://www.darpa.mil/grandchallenge04/sponsor_toolkit/congress_lang.pdf.

[13] Autonomous vehicles are "driverless vehicles," where a human does not need to be inside the vehicle to operate it.

[14] DARPA, DARPA Grand Challenge 2005 :Rules, October 8, 2004 at *http://www.darpa.mil /grandchallenge05/* Rules_8oct04.pdf.

[15] Personal communication, CRS with John Jennings, DARPA, on March 26, 2009.

[16] For more information, see http://www.darpa.mil/grandchallenge04/.

[17] For more information, see http://www.darpa.mil/grandchallenge05/.

[18] For more information, see http://www.darpa.mil/grandchallenge/index.asp.

[19] DARPA, *Prizes For Advanced Technology Achievements: Fiscal Year 2007 Annual Report*, January 2008 at http://www.darpa.mil/GRANDCHALLENGE/docs/DDRE_Prize_Report_FY07.pdf.

[20] Ibid.

[21] DARPA, *Prizes For Advanced Technology Achievements: Fiscal Year 2007 Annual Report*, January 2008 at http://www.darpa.mil/GRANDCHALLENGE/docs/DDRE_Prize_Report_FY07.pdf.

[22] For more information, see http://www.freedomprize.org/prizes/history.php.

[23] Personal communication, CRS with Karen Hanson, Executive Director, Freedom Prize, March 27, 2009..

[24] For more information on the H-prize, see http://www.hydrogenprize.com/indexNew.asp.

[25] Department of Energy, "Bright Tomorrow Lighting Competition," 73 *Federal Register* 35680, June 24, 2008.

[26] Personal communication, CRS with James Broderick, Department of Energy, June 10, 2009.

[27] Ibid.

[28] For a list of the current partners, see http://www.lightingprize.org/partners.stm.

[29] Progressive Automotive X Prize, "Prize Details," webpage at *http://www.progressiveau toxprize.org/prize-details.*

[30] Progressive Automotive X Prize, "FuelOurFutureNow.com Will Excite Students About Energy Efficiency," press release, February 3, 2009, at http://www.progressiveautoxprize.org/news-events/press-release/fuelourfuturenowcomwill-excite-students-about-energy-efficiency.

[31] For more information, see http://autoblog.xprize.org/axp/2009/02/progressive-automotive-x-prize-and-doe-launchfuelourfuturenowcom.html.

[32] For more details as to how green principles are evaluated in this competition, see http://www.americanlemans.com/images/sponsors/09_MICHELIN_GreenX_Challenge_ booklet.pdf.

[33] For more information regarding the Green Racing Initiative, see http://www.epa.gov/oms/ld-hwy/420f08031.htm. A fact sheet is available at http://www.epa.gov/oms/ld-hwy/420f08031.pdf.

[34] For more information regarding the Green Challenge Award, see http://www.epa.gov/oms/ld-hwy/420f08038.htm.

[35] GT vehicles generally include high-performance automobiles designed for long-distances.

[36] Robert Larson, Director Emeritus, Center for Transportation Research, Argonne National Laboratory, Presentation, April 15, 2009.

[37] For more information, see http://www.americanlemans.com/images 09_MICHELIN_Gree nX_Challenge_booklet.pdf.

[38] For more information, see http://www.americanlemans.com/index_green.php?sFile=12896.

[39] Ibid.

[40] For more information, see http://centennialchallenges.nasa.gov/.

[41] NASA, Innovative Partnership Programs, website at *http://centennialchallenges.nasa.gov/cc_challenges.htm#glove.*

[42] Tariq Malik, "Homemade Space Glove Wins NASA Contest," Space.com, May 4, 2007 at http://www.space.com/ businesstechnology/070504_astronaut_glove_win.html.

[43] For more information, see http://astronaut-glove.tripod.com/.

[44] For more information, see http://cafefoundation.org/v2/pav_home.php.

[45] For more information, see http://cafefoundation.org/v2/pav_gatchallenge.php.

[46] For the results, see http://cafefoundation.org/v2/pav_pavchallenge_2008results.php.

[47] For more information, see http://centennialchallenges.nasa.gov/, *http://cafefoundation.org/v2/cafenews_home.php,* and http://cafefoundation.org/v2/pdf_AGP/2009/AGP.TA.04.22.09.pdf/.

[48] Lunar regolith is the loose, fragmental material on the Moon's surface, commonly called "lunar soil." While regolith on the lunar surface is the product of meteoritic bombardment or rocks, regolith on Earth is due to weathering of rock. For more information, see http://www.spacegrant.hawaii.edu/class

[49] For more information, see http://regolith.csewi.org/.

[50] For more information, see http://space.xprize.org/lunar-lander-challenge.

[51] For more information, see http://www.spaceward.org/.

[52] For more information, see http://moonrox.csewi.org/.

[53] NASA FY2009 Budget Request, p. Cross-14 and p. 572.

[54] NASA FY2010 Budget Request, p. SUM-10. NASA is requesting $4.0 million for the program in FY2010. No funding was provided in FY2009. In H.Rept. 111-149, the House recommended not providing this funding due to affordability considerations.

[55] NASA FY2010 Budget Request, "Cross-agency support," p. Cross-42, at *http://www.nasa.gov/news/budget/* index.html.

[56] Ibid.

[57] Within the Department of Health and Human Services (HHS) is the Biomedical Advanced Research and Development Authority (BARDA), which is to provide an integrated, systematic approach to the development and purchase of the necessary vaccines, drugs, therapies, and diagnostic tools for public health medical emergencies. For more information on BARDA, see http://www.hhs.gov/aspr/barda/index.html.

[58] The information in this section is from CRS Report RL33907, Project BioShield: Appropriations, Acquisitions, and Policy Implementation Issues for Congress, by Frank Gottron.

[59] HHS, Office of the Assistant Secretary for Preparedness And Response, FY20 10 Online performance Appendix, p. 2 1-22 at *http://www.hhs.gov/asrt/ob/docbudget/2010aspropa.pdf.*

[60] Ibid.

[61] National Academy of Engineering, *Concerning Federally Sponsored Inducement Prizes in Engineering and Science* (Washington, DC: National Academy Press, 1999) at http://www.nap.edu/catalog.php?record_id=9724.

[62] Ibid.

[63] National Research Council, *Innovation Inducement Prizes at the National Science Foundation* (Washington, DC: National Academy Press, 2007).

[64] Richard G. Newell and Nathan E. Wilson, *Technology Prizes for Climate Change Mitigation*, RFF DP 05-3 3, Resources for the Future, June 2005 at http://www.rff.org/documents/RFF-DP-05-33.pdf.

[65] McKinsey & Company, *And the Winner Is ... Capturing the Promise of Philanthropic Prizes*, 2009 at http://www.mckinsey.com/clientservice/socialsector/And_the_winner_is.pdf.

[66] Thomas Kalil, *Prizes for Technological Innovation*, The Brookings Institution, December 2006 at http://www.brookings.edu/views/papers/200612kalil.pdf.

[67] Liam Brunt, Josh Lerner, and Tom Nicholas, *Inducement Prizes and Innovation,* CEPR Discussion Paper No. DP6917, July 2008 at http://ssrn.com/abstract=1307507.

[68] For more information, see CRS Report RL34559, *Intellectual Property in Industrial Designs: Issues in Innovation and Competition*, by John R. Thomas.

[69] For more information, see discussion of Federal Torts Claim Act in CRS Report RL34131, *Federal Liability for Flood Damage Related to Army Corps of Engineers Projects*, by Cynthia Brougher.

[70] Alex Schroeder, *The Application and Administration of Inducement Prizes in Technology*, IP-11-2004, Independence Institute, December 2004 at http://www.keionline.org/misc-docs/IP_11_2004.pdf.

[71] McKinsey & Company, *And the Winner is ... Capturing the Promise of Philanthropic Prizes*, 2009 at http://www.mckinsey.com/clientservice/socialsector/And_the_winner_is.pdf.

[72] Ibid.

[73] Adapted from National Research Council, Innovation Inducement Prizes at the National Science Foundation (Washington, DC: National Academy Press, 2007); National Academy of Engineering, Concerning Federally Sponsored Inducement Prizes in Engineering and Science (Washington, DC: National Academy Press, 1999); McKinsey & Company, And the Winner is ... Capturing the Promise of Philanthropic Prizes, 2009 at *http://www.mckinsey.com/clientservice/socialsector/And_the_winner_is.pdf.*

[74] For the most recent information on the status of these bills, see the Legislative Information System at http://www.congress.gov/.

[75] During the 110th Congress, several bills proposed prizes for greenhouse gas mitigation, farms that are net producers of both food and energy (H.R. 80), space and aeronautics (H.R. 4916), advanced technology achievements (S. 701), climate technology challenge (S. 280), and medical innovation (S. 2210), and innovation (S. 1371).

In: U.S. Innovation Programs
Editors: Allison E. Berger et al.
ISBN: 978-1-61122-809-0

Chapter 2

THE TECHNOLOGY INNOVATION PROGRAM

Wendy H. Schacht

SUMMARY

The Technology Innovation Program (TIP) at the National Institute of Standards and Technology (NIST) was established in 2007 to replace the Advanced Technology Program (ATP). This effort is designed "to support, promote, and accelerate innovation in the United States through high- risk, high-reward research in areas of critical national need," according to the authorizing legislation. Grants are provided to small and medium-sized firms for individual projects or joint ventures with other research organizations.

While similar to the Advanced Technology Program in the promotion of R&D that is expected to be of broad-based economic benefit to the nation, TIP appears to have been structured to avoid what was seen as government funding of large firms that opponents argued did not necessarily need federal support for research. The committee report to accompany H.R. 1868, part of which was incorporated into the final legislation, stated that TIP replaces ATP in consideration of a changing global innovation environment focusing on small and medium-sized companies. The design of the program also "acknowledges the important role universities play in the innovation cycle by allowing universities to fully participate in the program."

The elimination of ATP and the creation of TIP have renewed the debate over the role of the federal government in promoting commercial technology development. In arguing for less direct federal involvement, advocates of this approach believe that the market is superior to government in deciding technologies worthy of investment. Mechanisms that enhance the market's opportunities and abilities to make such choices are preferred. It is suggested that agency discretion in selecting one technology over another can lead to political intrusion and industry dependency. On the other hand, supporters of direct methods argue that it is important to focus on those technologies that have the greatest promise as determined by industry and supported by matching funds from the private sector. They assert that the government can serve as a catalyst for cooperation. As the 111th Congress continues to make budget decisions, the discussion may serve to redefine thinking about governmental efforts in facilitating technological advancement in the private sector.

INTRODUCTION

The Technology Innovation Program (TIP) at the National Institute of Standards and Technology (NIST) was created to "to support, promote, and accelerate innovation in the United States through high-risk, high-reward research in areas of critical national need," according to the authorizing legislation. The intent of the program is to provide grants to small and medium-sized firms for individual projects or joint ventures with other research organizations to undertake work that

(A) has the potential for yielding transformational results with far-ranging or wide-ranging implications;
(B) addresses critical national needs within the National Institute of Standards and Technology's areas of technical competence; and
(C) is too novel or spans too diverse a range of disciplines to fare well in the traditional peer- review process.[1]

NIST published the final rule prescribing the policies and procedures for the TIP activity on June 25, 2008 (15 C.F.R. Part 296). Small or medium-sized for-profit firms are eligible for individual project awards of up to $3 million over three years. Collaborative research ventures including small or medium-sized companies, national laboratories, universities, or other non-profit research institutions may be funded for a total of up to $9 million over five years. A competitive, merit- based process is to be used to make grants of up to 50% of total project costs. In January 2009, nine awards were announced for "new research projects to develop advanced sensing technologies that would enable timely and detailed monitoring and inspection of the structural health of bridges, roadways and water systems that comprise a significant component of the nation's public infrastructure." According to TIP, $42.5 million in federal money was expected to be matched by $45.7 in private sector support. Twenty more awards were announced in December 2009 totaling almost $71.0 million in NIST financing with approximately $145.7 million in funding from other sources. Of the projects selected for the two solicitations, thirteen were in the area of monitoring and inspection of civil infrastructure; four were in the area of advanced repair of civil infrastructure; eleven were in the area of process scale up for advanced materials; and one was in the area of predictive modeling for advanced materials.

FUNDING

The Technology Innovation Program was authorized by the America COMPETES Act (P.L. 110- 69) and replaced the Advanced Technology Program (ATP). The FY2008 Consolidated Appropriations Act, P.L. 110-161, provided the initiative $65.2 million (with an additional $5 million in unobligated balances from the FY2007 ATP appropriation). This was 17.6% less than FY2007 funding for ATP. According to NIST, the major portion of FY2008 support was to be used to meet previous ATP funding commitments.

The FY2009 budget request did not contain any financial support for TIP. During the 110th Congress, no final FY2009 appropriations legislation was enacted for the program; P.L. 110-329, the Consolidated Security, Disaster Assistance, and Continuing Appropriations Act, 2009, funded TIP at FY2008 levels through March 6, 2009. FY2009 funding of $65.0 million was subsequently provided by P.L. 111-8, the FY2009 Omnibus Appropriations Act.

For FY2010, $69.9 million was appropriated for TIP by P.L. 111-117, the Consolidated Appropriations Act, 2010, an increase of 7.5% over the previous fiscal year. The figure reflected that amount included in President Obama's budget request and H.R. 2847 as originally passed by both the House and the Senate.

The Administration's FY2011 budget includes $79.9 million for TIP, a 14.3% increase in support over FY2010.

BACKGROUND

The Technology Innovation Program replaced the Advanced Technology Program[2] which was established by Title V of the Omnibus Trade and Competitiveness Act (P.L. 100-418). ATP was intended "to serve as a focal point for cooperation between the public and private sectors in the development of industrial technology" and to help solve "problems of concern to large segments of an industry," as noted in the conference report to accompany the bill. Located within NIST, in recognition of the laboratory's ongoing relationship with industry, ATP provided seed funding to single companies or to industry-led consortia of universities, businesses, and/or government laboratories for development of generic (broad-based), pre-competitive technologies that had applications across industries. Awards, based on technical and business merit, were for high-risk work past the basic research stage but not yet ready for commercialization. Market potential was an important consideration in project selection.

Awards were for either product or process technology development. Individual firms were restricted to funding of $2 million over three years. Money was to be used only for direct R&D costs. Large firms provided at least 60% of total (direct and indirect) projects costs; small and medium-sized companies were not required to cost-share direct costs. Joint ventures could receive up to five years of financing for any amount, limited only by availability. In such cases, the private sector provided more than 50% of funding. While universities and federal laboratories could participate in collaborative work, the ATP grant was made solely to companies.

According to NIST, through the end of 2007 when ATP was terminated, 824 projects had been funded, of which about 28% were joint ventures. Approximately $1.6 billion in federal funds have been matched by $1.5 billion from the private sector. Small businesses or cooperative efforts led by such firms made up almost 68% of the awardees. (The legislation creating TIP required that the Director of NIST "continue to provide support originally awarded under the Advanced Technology Program, in accordance with the terms of the original award and consistent with the goals of the Technology Innovation Program.")

Financing for ATP was a continuing issue in the appropriations debate prior to the elimination of the program. Opponents of ATP cited it as a prime example of "corporate welfare," whereby the federal government invested in applied research activities that, they emphasized, should be conducted by the private sector. Others defended ATP, arguing that it assisted businesses (and small manufacturers) in developing technologies that, while crucial to industrial competitiveness, would not or could not be developed by the private sector alone.

TIP appears to have been designed to avoid what opponents of the ATP program argued was government funding of large firms that had no need for federal support for R&D activities. The House Committee on Science and Technology report to accompany H.R. 1868, part of which was incorporated into the COMPETES Act, stated that TIP replaces ATP in consideration of a changing global innovation environment focusing on small and medium-sized companies. The structure of TIP also "acknowledges the important role universities play in the innovation cycle by allowing universities to fully participate in the program."[3]

A DIFFERENT APPROACH

While similar to ATP in the promotion of high-risk R&D that would be of broad-based economic benefit to the Nation, there are several differences in the operation of TIP primarily associated with the size of eligible companies. Financial support under TIP is limited to small and medium- sized businesses whereas grants under ATP were available to companies regardless of size. In addition, the Advanced Technology Program required that joint ventures include two separately owned for-profit firms and could include universities, government laboratories, and other research establishments as participants in the project but not as recipients of the grant. In the TIP initiative, a joint venture may involve two separately owned for-profit companies but may also be comprised of one small or medium-sized firm and a university (or other non-profit research institution). A single company could receive up to $2 million for up to three years under ATP; under TIP, the participating company (which must be a small or medium-sized business) may receive up to $3 million over three years. In ATP, small and medium-sized companies were not required to cost share (large firms provided 60% of the total cost of the project) while in TIP there is a 50% cost sharing requirement which, again, only applies to the small and medium-sized businesses that are eligible. There were no funding limits for the five-year funding available for joint ventures under ATP; the TIP limits joint venture funding to $9 million for up to five years. The Advisory Board that was created to assist in the Advanced Technology Program included industry representatives as well as federal government personnel and representatives from other research organizations. The Advisory Board for the Technology Innovation Program would be comprised of only private sector members.

Awards are made for work in areas of "Critical National Need" defined by NIST as:

> An area that justifies government attention because the magnitude of the problem is large and the societal challenges that need to be overcome are not being addressed, but could be addressed through high-risk, high-reward research.[4]

To identify the needs for which TIP awards grants, the organization solicits input from government agencies, advisory groups, industry organizations, universities, and NIST.[5] These are further refined by TIP working groups and specific topics are selected. Current areas include civil infrastructure and manufacturing while future areas of interest include advanced robotics, complex networks, energy, healthcare, sustainability, and water.[6]

ISSUES AND OBSERVATIONS

The effort to terminate the Advanced Technology Program, along with additional attempts to withdraw government support for certain other technology development efforts, appeared to reflect a philosophy that eschewed direct federal financing of private sector R&D efforts aimed at the commercialization of new technologies and production processes. Such activities are seen by opponents as "industrial policy," the means by which government rather than the marketplace "picks winners and losers." Instead, measures that would occasion a better investment environment for industry to expand their innovation-related efforts would, proponents argue, be preferable to government funding.

The current approach, including the new Technology Innovation Program, involves varied mechanisms to facilitate technological advancement.

Legislation has created a body of laws, programs, and policies that involve both indirect and direct measures to stimulate technology advancement in the private sector. Indirect incentives include a research and experimentation tax credit; changes to the antitrust laws to encourage collaborative R&D and cooperative manufacturing ventures; alterations of patent ownership policies to facilitate government industry-university interaction; and practices to promote technology transfer. Direct measures involve, among other things, federal funding for TIP and the Small Business Innovation Research Program. These cost-shared programs have been supported, in part, because of what proponents argue is their potential contribution to the country's national or economic security.

The elimination of ATP and the creation of TIP have renewed the debate over the role of the federal government in promoting commercial technology development. In arguing for less direct federal involvement, advocates believe that the market is superior to government in deciding technologies worthy of investment. Mechanisms that enhance the market's opportunities and abilities to make such choices are preferred. It is suggested that agency discretion in selecting one technology over another can lead to political intrusion and industry dependency. On the other hand, supporters of direct methods argue that it is important to focus on those technologies that have the greatest promise as determined by industry and supported by matching funds from the private sector. They assert that the government can serve as a catalyst for cooperation.

Technological progress is important to the nation because of its contribution to economic growth and a high standard of living. How best to achieve this continues to be debated. Critics viewed ATP as a means for a federal agency to select commercial firms and/or technologies for support. They maintained that the absence of market-generated decisions will result in technologies that can not be utilized productively by participating companies. Such a program, they argued, encourages selection of well-written proposals rather than assistance for truly important technologies. However, proponents stressed that ATP was market driven and that the technical areas for investment had been developed in conjunction with industry. In addition, companies were required to put up significant amounts of funding and survive a rigorous business review; procedures that made the ATP different from other federal efforts. Replacing ATP with the Technology Innovation Program may be one response to criticisms that large firms should not be recipients of this form of federal research funding, support that should be reserved for small and medium-sized companies which do not have the financial resources available to major corporations.

End Notes

[1] P.L. 110-69

[2] For additional information on ATP see CRS Report 95-3 6, *The Advanced Technology Program*, by Wendy H. Schacht.

[3] U. S. Congress, House Committee on Science and Technology, *Technology Innovation and Manufacturing Stimulation Act of 2007*, Report to accompany H.R. 1868, H.Rept. 110-115, April 30, 2007, 21.

[4] Slides provided by Lorel Wisniewski, Deputy Director, Technology Innovation Program, NIST, May 19, 2010, 2.

[5] National Institute of Standards and Technology, *Technology Innovation Program, Transforming America's Future Through Innovation, Annual Report 2008*, December 2009, 4.

[6] *Slides,* 4.

In: U.S. Innovation Programs
Editors: Allison E. Berger et al.
ISBN: 978-1-61122-809-0

Chapter 3

SMALL BUSINESS INNOVATION RESEARCH (SBIR) PROGRAM

Wendy H. Schacht

SUMMARY

In 1982, the Small Business Innovation Development Act (P.L. 97-2 19) established Small Business Innovation Research (SBIR) programs within the major federal research and development (R&D) agencies designed to increase participation of small innovative companies in federally funded R&D. Government agencies with R&D budgets of $100 million or more are required to set aside a portion of these funds to finance the SBIR activity. Through FY2007, over $22.3 billion in awards have been made for more than 100,016 projects. Extended several times, the program was scheduled to sunset on September 30, 2008. Although, to date, no specific legislation has reauthorized the program, the Small Business Administration determined that P.L. 110-235 temporarily extended the SBIR activity through March 20, 2009. Subsequently, P.L. 111- 10 provided an additional extension of the program through July 31, 2009; P.L. 111-43 extended it through September 30, 2009; and P.L. 111-66 extended the effort through October 31, 2009. The SBIR program was once again extended through April 30, 2010, by P.L. 111-89 and through September 30, 2010, by P.L. 111-214. On July 8, 2009, H.R. 2965, a bill to reauthorize and alter the SBIR initiative, passed the House; the Senate substituted the language of S. 1233 (amended) and passed its version of H.R.

2965 on July 13, 2009. For further information on SBIR reauthorization activity see CRS Report RS22865, *The Small Business Innovation Research (SBIR) Program: Reauthorization Efforts*, by Wendy H. Schacht.

Program Description

The Small Business Innovation Research (SBIR) program is designed to increase the participation of small, high technology firms in the federal R&D endeavor. Congressional support for the initiative was predicated upon the belief that while technology-based companies under 500 employees tended to be highly innovative, and innovation is essential to the economic well-being of the United States, these businesses were underrepresented in government R&D activities. Agency SBIR programs guarantee this sector a portion of the government's R&D budget to compensate for what was viewed as a preference for contracting with large firms.

Current law requires that every federal department with an R&D budget of $100 million or more establish and operate an SBIR program. A set percentage of that agency's applicable extramural research and development budget—originally at 1.25%, now at 2.5%—is to be used to support mission-related work in small companies.

The objectives of the SBIR program include stimulation of technological innovation in the small business sector, increased use of this community to meet the government's R&D needs, additional involvement of minority and disadvantaged individuals in the process, and expanded commercialization of the results of federally funded R&D. To achieve this, agency SBIR efforts involve a three-phase activity. In the first phase, awards up to $100,000 (for six months) are provided to evaluate a concept's scientific or technical merit and feasibility. The project must be of interest to and coincide with the mission of the supporting organization. Projects that demonstrate potential after the initial endeavor may compete for Phase II awards of up to $750,000 (lasting one to two years) to perform the principal R&D. Phase III funding, directed at the commercialization of the product or process, is expected to be generated in the private sector. Federal dollars may be used if the government perceives that the final technology or technique will meet public needs. P.L. 102-564 directed agencies to weigh commercial potential as an additional factor in evaluating SBIR proposals.

As of FY2009, 11 departments have SBIR programs including the Departments of Agriculture, Commerce, Defense (DOD), Education, Energy, Health and Human Services, Homeland Security, and Transportation; the Environmental Protection Agency; the National Aeronautics and Space Administration (NASA); and the National Science Foundation (NSF). Each agency's SBIR activity reflects that organization's management style. Individual departments select R&D interests, administer program operations, and control financial support. Funding can be disbursed in the form of contracts, grants, or cooperative agreements. Separate agency solicitations are issued at established times.

The Small Business Administration (SBA) created broad policy and guidelines under which individual departments operate SBIR programs. The agency monitors and reports to Congress on the conduct of the separate departmental activities. Criteria for eligibility in the SBIR program include companies that are independently owned and operated; not dominant in the field of research proposed; for profit; the employer of 500 or less people; the primary employer of the principal investigator; and at least 51% owned by one or more U.S. citizens or lawfully admitted permanent resident aliens. A rule change, effective January 3, 2005, permits subsidiaries of SBIReligible companies to participate as long as the parent company meets all SBIR requirements.

A pilot effort to encourage commercialization of university and federal laboratory R&D by small companies was created by P.L. 102-564 and reauthorized several times through FY2009. The Small Business Technology Transfer program (STTR) provides funding for research proposals that are developed and executed cooperatively between a small firm and a scientist in a research organization and fall under the mission requirements of the federal funding agency. Up to $100,000 in Phase I financing is available for one year; Phase II awards of up to $750,000 may be made for two years. Currently funded by a set-aside of 0.3% of the extramural R&D budget of departments that spend over $1 billion per year on this effort, the Departments of Energy, Defense, and Health and Human Services, NASA, and NSF participate in the STTR program.

IMPLEMENTATION

The Government Accountability Office (GAO; formerly the General Accounting Office) is legislatively directed to assess the implementation of the Small Business Innovation Development Act, as amended, and has issued a series of reports documenting its findings. A 1987 study found that both the evaluation and selection processes were sufficient to "reasonably" insure awards were based on technical merit. It was also determined that the majority of agencies were not awarding Phase I grants and contracts within the six-month time frame required by the SBA guidelines. Another GAO report the following month surveyed the participants and noted that most were "generally satisfied" with the administration of SBIR programs.

In 1989, GAO reported that agency heads found the SBIR effort to be beneficial and met the organization's R&D needs. Most indicated that the "SBIR programs had developed new research areas, placed more emphasis on the application of research results, and led to wider use of small businesses as research performers." The study concluded that projects were, for the most part, of high quality. At DOD and NASA, however, SBIR efforts stressed R&D to meet agency mission requirements in contrast to other SBIR programs that focused on commercialization for private sector markets. All of the departments stated that SBIR projects, when compared with other research activities, had greater potential to result in new products and processes.

Testimony presented by GAO in 1991 stated that the program "clearly is doing what Congress asked it to do in achieving commercial sales and developmental funding from the private sector." An SBA study found that approximately one in four SBIR projects will result in the sale of new commercial products or processes. Another GAO report issued in May 1992 noted that despite a short time frame and the fact that many SBIR projects had not had sufficient time to mature into marketable technologies and techniques, "the program is showing success in Phase III activity." As of July 1991, almost two-thirds of the projects already had sales or received additional funding (primarily from the private sector) totaling approximately $1.1 billion.

The 1992 study also identified several issues for possible further congressional exploration. According to GAO, DOD placed less emphasis on commercialization than other agencies and utilized the SBIR program primarily to address the department's R&D needs. Questions were raised about the requirements for competitive bidding when companies looked to federal departments for Phase III contracts after successfully completing Phases I and II. GAO noted that clarification of the Competition in

Contracting Act of 1984 (as amended) might be necessary. In addition, there was disagreement over whether the federal agency or the small firm should continue to work on technology development after the cessation of SBIR project funding. GAO also concluded that firms receiving multiple Phase II awards tended to have lower Phase III sales and less additional developmental support. The reasons for this remained unclear, but the suggestion was made that these companies may have focused on securing funds through SBIR awards rather than through commercialization of their R&D results.

A March 1995 GAO report found that multiple Phase II funding had become a problem, particularly at NSF, NASA, and DOD. Among the reasons cited were the failure of companies to identify identical proposals made elsewhere in violation of the mandatory certification procedure; uncertainty in definitions and guidelines concerning "similar" research; and lack of interagency mechanisms to exchange information on projects. Several recommendations were made to address duplication. GAO testimony presented in March 1996 indicated that the SBA had taken steps to implement these suggestions. The study also determined that the quality of research appeared to have "kept pace" with the program's expansion, although it was still too early to make a definitive judgment. Factors supporting this assessment included the substantive level of competition, more proposals deemed meritorious than could be funded by agencies, and appraisals by departmental SBIR personnel indicating the high quality of submissions.

Another GAO study, released in April 1998, noted that between 35% and 50% of SBIR projects had resulted in sales or additional private sector investment. Despite earlier indications of problems associated with multiple award winners, this chapter found that such firms have similar commercialization rates as single awardees. Critical technology lists were being used to determine agency solicitations and there was little evidence of participation by foreign firms. While several agencies had new programs to assure continuity in funding, there were indications of possible inaccuracies in defining the extramural R&D budgets upon which the set-aside is based.

The June 1999 GAO analysis reported that SBIR awards tend to be concentrated both geographically and by firm despite widespread participation in the program. "The 25 most frequent winners, which represent fewer than 1 percent of the companies in the program, received about 11 percent of the program's awards from fiscal year 1983 through fiscal year 1997." Businesses in a small number of states, particularly California and Massachusetts, were awarded the most number of projects. The study also noted that while commercial potential is considered by all agencies, each has developed different evaluation approaches. Other goals, including innovation and responsiveness to agency mission, still remain important in determining awards.

A more recent report by GAO (June 2005) found that it is still difficult to adequately "assess the performance of the SBIR program" although the effort appears to be achieving its goal of "enhanced" participation of small business in the R&D enterprise. Utilizing "commercialization" as a measure may not be sufficient because other agency goals were being met such as research needs or expanded innovation. Success in the commercial market did not take into account the R&D requirements of departments like DOD or NASA. In a report the following year (October 2006), GAO noted that the agencies reporting to the SBA did not always provide the necessary data in the format required by SBA. GAO concluded that the "agencies need to strengthen [their] efforts to improve the completeness, consistency, and accuracy of awards data."

GAO also has evaluated the STTR program. A January 1996 report found that, in general, federal agencies favorably rated the quality of winning proposals (in the first year) and that most projects had commercial potential, although the costs might be high. The government had taken steps to avoid potential conflicts of interest between federal laboratories and departmental headquarters. There was no indication that this pilot effort was competing for proposals with the established SBIR activity or "reducing the quality of the agencies' R&D in general." Instead it was credited for encouraging collaborative work. Yet, GAO noted that because the programs are so similar, there are questions whether or not a separate activity is necessary. Any real evaluation of success in technology transfer, however, could not be accomplished for several years because of the time needed to bring the results of R&D to the commercial marketplace. These findings were reiterated in testimony given by GAO in May and September 1997.

A June 2001 GAO study of all companies which received STTR awards between FY1995 and FY1 997 noted the participant's belief that both the firms and the research institutions contributed to expanded R&D although the private sector was more influential in determining the direction of the research. The companies "reported about $132 million in total sales and about $53 million in additional developmental funding." They identified 41 new patents and the creation of 12 new spin-off firms. Further, the awardees preferred that the STTR program remain separate from the SBIR activity.

Table 1. SBIR Program: Dollars Awarded and Projects Funded

Fiscal Year	Dollars Awarded (millions)			Awards		
	Phase I	Phase II	Total	Phase I	Phase II	Total[a]
FY1 983	44.5	—	44.5	686	—	686
FY1984	48.0	60.4	108.4	999	338	1,337
FY1 985	69.1	130.0	199.1	1,397	407	1,804
FY1 986	98.5	199.4	297.9	1,945	564	2,509
FY1 987	109.6	240.9	350.5	2,189	768	2,957
FY1988	101.9	284.9	389.1[a]	2,013	711	2,724
FY1989	107.7	321.7	431.9[a]	2,137	749	2,886
FY1990	118.1	341.8	460.7[a]	2,346	837	3,183
FY1991	127.9	335.9	483.1[a]	2,553	788	3,341
FY1 992	127.9	371.2	508.4[a]	2,559	916	3,475
FY1 993	154.0	490.7	698.0[a]	2,898	1,141	4,039
FY1 994	220.4	473.6	71 7.6[a]	3,102	928	4,030
FY1 995	232.1	601.9	834.1 [a]	3,085	1,263	4,348
FY1 996	228.9	645.8	91 6.3[a]	2,841	1,191	4,032
FY1 997	277.6	789.1	1,1 06.7[a]	3,371	1,404	4,775
FY1 998	262.3	804.4	1,066.7	3,022	1,320	4,342
FY1999	299.5	797.0	1,096.5	3,334	1,256	4,590
FY2000	302.0	888.2[b]	1,190.2	3,166	1,330	4,496
FY2001	317.1	977.3	1,294.4	3,215	1,533	4,748
FY2002	411.5	1 ,023.4[b]	1,434.9	4,243	1,577	5,820
FY2003	445.4	1,214.7	1,660.1	4,465	1,759	6,224
FY2004	498.7	1,368.7	1,867.4	4,638	2,013	6,651
FY2005	461.2	1,404.7	1,865.9	4,300	1,871	6,171
FY2006	411.2	1,472.0	1,883.2	3,836	2,026	5,862
FY2007	447.3	1,197.4	1,644.7	3,814	1,542	5,356

Source: Small Business Administration Data.

a. Includes modifications to previous awards and funds set aside for proposals in negotiation.

b. Dollars obligated can include modifications to previous year's awards.

AWARDS

From its inception in FY1983 through FY2007, over 100,016 awards have been made totaling more than $22.3 billion. **Table 1** summarizes the funding and the number of projects selected for the SBIR program as provided by the

SBA; information on the STTR program is contained in **Table 2**. (Figures for FY2008 are not yet available from the Small Business Administration.)

ISSUES FOR CONSIDERATION

The SBIR program was scheduled to sunset on September 30, 2008. Although, to date, no specific legislation has reauthorized the program, the Small Business Administration determined that P.L. 110-235 temporarily extended the SBIR activity through March 20, 2009. Subsequently, P.L. 111-10 provided an additional extension of the program through July 31, 2009; P.L. 111-43 extended it through September 30, 2009; and P.L. 111-66 extended the effort through October 31, 2009. The SBIR program was then extended through April 30, 2010, by P.L. 111-89 and through September 30, 2010 by P.L. 111-214. On July 8, 2009, H.R. 2965, a bill to reauthorize and alter the SBIR initiative, passed the House; the Senate substituted the language of S. 1233 (amended) and passed its version of H.R. 2965 on July 13, 2009.

Table 2. STTR Program: Dollars Awarded and Projects Funded

Fiscal Year	Dollars Awarded (millions)			Awards		
	Phase I	Phase II	Total	Phase I	Phase II	Total
FY1994	18.9	—	18.9	198	—	198
FY1995	23	10.7	33.7	238	22	260
FY1996	22.7	41.8	64.5	238	88	326
FY1997	24.2	44.9	69.1	260	89	349
FY1998	19.7	45.1	64.8	208	109	317
FY1999	24.3	40.6	64.9	251	78	329
FY2000	23.9	45.9	69.8	233	95	328
FY2001	24.2	53.2	77.4	224	113	337
FY2002	36.4	55.4	91.8	356	114	470
FY2003	41.1	50.7	91.8	397	111	508
FY2004	79.7	110.3	190	674	195	869
FY2005	73.9	146.4	220.3	611	221	832
FY2006	74.0	152.3	226.3	644	234	878
FY2007	83.5	159.4	242.9	634	213	847

Source: Small Business Administration data.

Certain issues might be considered if the program is to be reauthorized. (For further information on SBIR reauthorization activity see CRS Report RS22865, *The Small Business Innovation Research (SBIR) Program: Reauthorization Efforts*, by Wendy H. Schacht.) Much of the current debate revolves around the existing regulations that require at least 51% ownership by an individual or individuals. Some experts argue participation by small firms that are majority- owned by venture capital companies should be permitted. Proponents of this change maintain that, particularly in the biotechnology sector, the most innovative companies are not able to use the SBIR program because they do not meet these ownership criteria. Opponents of altering the eligibility requirements argue that the program is designed to provide financial assistance where venture capital is not available. They assert that the program's objective is to bring new concepts to the point where private sector investment is feasible.

An additional concern is the extent to which program participants are mandated to report activities and results. P.L. 106-554 placed added requirements on companies to provide information; it remains to be determined if these requirements have been successfully implemented. Other issues that might be addressed include whether the problems identified by GAO associated with the duplication of awards has been adequately resolved. Are the SBIR and STTR programs meeting their different mandated objectives or are they serving an identical purpose? Does the focus on commercialization raise concerns by those who argue that the government has no role in directly supporting industrial research and development? These and other questions may be explored as the 111th Congress considers reauthorization of the Small Business Innovation Research program.

In: U.S. Innovation Programs
Editors: Allison E. Berger et al.

ISBN: 978-1-61122-809-0

Chapter 4

The Small Business Innovation Research (SBIR) Program: Reauthorization Efforts

Wendy H. Schacht

Summary

The Small Business Innovation Development Act of 1982, P.L. 97-219, created Small Business Innovation Research (SBIR) programs within the major federal research and development (R&D) agencies. This effort was intended to increase participation of small innovative companies in federally funded R&D. Government agencies with extramural R&D budgets of $100 million or more are required to set aside a portion of these funds to support research and development in small firms through the SBIR program. The program was extended several times and was scheduled to terminate on September 30, 2008. Although, to date, no specific legislation reauthorized the program, the Small Business Administration determined that P.L. 110-235 temporarily extended the SBIR activity through March 20, 2009. P.L. 111-10 provided an additional extension of the program through July 31, 2009; P.L. 111-43 extended it through September 30, 2009; and P.L. 111-66 extended the effort through October 31, 2009. P.L. 111-89 once again extended the SBIR program through April 30, 2010, after which P.L. 111-214 extended the effort through September 30, 2010. H.R. 2965, a bill to reauthorize and alter the

SBIR initiative, passed the House after which the Senate substituted the language of S. 1233 (amended) and approved its version of H.R. 2965.

THE CURRENT PROGRAM

Congress has demonstrated an ongoing interest in the small business sector. Addressing issues related to economic growth and competitiveness, special consideration has been given to small, high-tech firms for several reasons, including some data that indicate such companies tend to be highly innovative, play a significant role in technological advancement, and contribute to the high standard of living in the United States. Such was the rationale behind legislation creating the Small Business Innovation Research (SBIR) program, an effort to increase that portion of the federal research and development (R&D) budget provided to small enterprises for work associated with the mission responsibilities of government departments and agencies. Believing that small companies were underrepresented in government R&D activities, P.L. 97-219, as amended, established agency SBIR programs to guarantee this sector a portion of the government's research and development budget to compensate for what was viewed as a federal contracting preference for large corporations.

Current law requires that every federal department with an extramural R&D budget of $100 million or more establish and operate an SBIR program.[1] A set percentage of that agency's extramural research and development budget[2]—originally at 1.25%, now at 2.5%—is to be used to support mission-related work in small companies. To be eligible to compete in the program, a company must be independently owned and operated; not dominant in the field of research proposed; for profit; the employer of 500 or fewer people; the primary employer of the principal investigator; and at least 51% owned by one or more U.S. citizens or lawfully admitted permanent resident aliens. A rule change, effective January 3, 2005, permits subsidiaries of SBIR-eligible companies to participate as long as the parent company meets all SBIR requirements.

Agency SBIR efforts involve a three-phase activity. In the first phase, awards up to $100,000 (for six months) are provided to evaluate a concept's scientific or technical merit and feasibility. The project must be of interest to and coincide with the mission of the supporting organization. Projects that demonstrate potential after the initial endeavor may compete for Phase II awards of up to $750,000 (lasting one to two years) to perform the principal R&D. Phase III funding, directed at the commercialization of the product or process, is expected to be generated in the private sector. Federal dollars, but not SBIR funds, may be used if the government perceives that the final technology or technique will meet public needs. P.L. 102-564 directed agencies to weigh commercial potential as an additional factor in evaluating SBIR proposals.

As of FY2008, 11 departments have SBIR programs, including the Departments of Agriculture, Commerce, Defense (DOD), Education, Energy, Health and Human Services, Homeland Security, and Transportation; the Environmental Protection Agency; the National Aeronautics and Space Administration (NASA); and the National Science Foundation (NSF). Each agency's SBIR activity reflects that organization's management style. Individual departments select R&D interests, administer program operations, and control financial support. Funding may be disbursed in the form of contracts, grants, or cooperative agreements. Separate agency solicitations are issued at established times.

The Small Business Administration (SBA) created broad policy and guidelines under which individual departments operate SBIR programs. The agency monitors and reports to Congress on the conduct of the separate departmental activities.

A pilot effort to encourage commercialization of university and federal laboratory R&D by small companies was created by P.L. 102-564 and reauthorized several times through FY2009. The Small Business Technology Transfer program (STTR) provides funding for research proposals that are developed and executed cooperatively between a small firm and a scientist in a research organization and fall under the mission requirements of the federal funding agency. Up to $100,000 in Phase I financing is available for one year; Phase II awards of up to $750,000 may be made for two years. Currently funded by a set-aside of 0.3% of the extramural R&D budget of departments that spend over $1 billion per year on this effort, the Departments of Energy, Defense, and Health and Human Services, NASA, and NSF participate in the STTR program.

Table 1. 111th Congress: Comparison of H.R. 2965 (as passed by House and Senate) by Selected Issue

Issue	H.R. 2965 Passed by House	H.R. 2965 Passed by Senate
Reauthorization	The SBIR and STTR programs are extended through September 30, 2011.	The bill extends the SBIR and STTR programs through September 30, 2017.
Award Levels	Phase I SBIR and STTR awards are increased from \$100,000 to \$250,000 and Phase II SBIR and STTR awards are increased from \$750,000 to \$2,000,000. Increases the maximum allowable Phase I award to \$300,000 and Phase II award to \$2,250,000 for veteran-owned and controlled small businesses.	Phase I SBIR and STTR awards are increased from \$100,000 to \$150,000 and Phase II SBIR and STTR awards are increased from \$750,000 to \$1,000,000.
SBIR Set-Aside Amounts	No similar provision.	The set-aside is increased from 2.5% in 2009 and 2010 by 0.1% per year up to 3.5% in 2020 and beyond with the following exception: the increases in funding generated by the set-aside in excess of 2.5% at the Department of Defense and the Department of Energy are not to be used for new Phase I or II awards but for "activities that further the technology readiness levels of technologies being developed under Phase II awards. "
STTR Set-Aside Amounts	No similar provision.	The set-aside is increased from 0.3% to 0.4% for 2011 and 2012; to 0.5% for 2013 and 2014; and to 0.6% from 2015 on.
Sequential Awards	Small business that receives one Phase II award is eligible to receive additional Phase II awards. A Phase II award can be made only after a Phase I award has been completed or the small business has met	A recipient of a Phase I grant from one federal agency would be permitted to apply for a Phase II award from another agency to pursue the original work. A small business would be allowed to

Table 1. (Continued)

Issue	H.R. 2965 Passed by House	H.R. 2965 Passed by Senate
	the requirements set forth in 15 U.S.C. 638 subsection (e)(4)(A) re technical merit and commercial potential. Allows veteran-owned and controlled small businesses to bypass Phase I and go directly to Phase II.	switch between the SBIR and STTR programs.
Grant Award Priority	Priority in granting awards is to be given to small companies owned by veterans and those in areas that have lost a major source of employment	Phase III awards are to go to companies that developed the technologies in Phase I and Phase II. Grants are to be made on a competitive and merit basis.
Majority Venture Capital Owned Small Businesses	For the SBIR and STTR programs, the bill would allow majority venture capital ownership in a small business if not more than 50% of the firm is owned by one venture capital company and the employees of the venture capital company are not a majority of the small firm's board of directors. If the venture capital company is controlled by a business with more than 500 employees, the small business is eligible only if not more than two large venture capital companies have ownership interest in the small firm and these large venture capital companies do not collectively own more than 20% of the small business.	The bill permits NIH to award not more than 18% of SBIR funds to majority venture capital-owned small businesses so long as "no single venture capital company owns more than 49 percent of the small business concern" upon a written determination provided to the Administrator of the SBA and the Senate Committee on Small Business and Entrepreneurship and the House Committee on Small Business. Allows other participating agencies to award not more than 8% of SBIR funds to majority venture capital-owned small businesses so long as "no single venture capital company owns more than 49 percent of the small business concern" upon a written determination provided to the Administrator of the SBA and the Senate Committee on Small Business and Entrepreneurship and the House Committee on Small Business.

Table 1. (Continued)

Issue	H.R. 2965 Passed by House	H.R. 2965 Passed by Senate
Outreach	The bill authorizes the SBA to support outreach and other efforts to increase participation in the SBIR program, with particular emphasis on small firms in rural areas. $10 million in funding is authorized for this activity in FY2010 and FY2011. Programs receiving outreach grants are to direct their activities to underperforming states and regions, women, service-disabled veterans, Native American-owned businesses, or minority-owned organizations, as well as to areas of high unemployment. A program of matching grants to minority institutions is created to encourage minority participation. In FY2010 and FY2011, $4 million per year is authorized to be appropriated for this effort.	The bill authorizes and makes changes to the Federal and State Technology Partnership (FAST) program which provides grants to organizations to provide outreach designed to encourage increased participation in the SBIR program. Rural outreach and outreach to states with low program participation are emphasized. Creates an SBIR-STEM Workforce Development Grant Pilot Program. Provides for the authorization of $1 million per year from 2010 through 2015 for this activity.
Commercialization Activities	Each SBIR agency is to create a commercialization program to facilitate "progress" to Phase III. The program may include partnerships, mentoring with prime contractors, transition assistance, and multiple Phase II awards, among other things. Funding for $27.5 million is authorized for FY2010 and after for the SBA to transfer to the agencies for commercialization programs. The bill requires agencies to establish procedures to	The bill provides for commercialization pilot programs for Phase II SBIR and STTR technologies in DOD and the civilian agencies. Encourages SBIR and STTR awards to small businesses that work with federal laboratories or are involved in cooperative research and development agreements (CRADAs).

Table 1. (Continued)

Issue	H.R. 2965 Passed by House	H.R. 2965 Passed by Senate
	encourage SBIR awardees to develop partnerships with other organizations, including prime contractors, business incubators, venture capital companies, and large business, to assist them in moving to Phase III. Mandates commercialization efforts to assist small companies affected by the termination of the shuttle program.	
Areas of Concentration	Agencies are directed to focus on research in energy, rare diseases, transportation, nanotechnology, medical technology, water delivery systems, and renewable energy.	Agencies are encouraged to make SBIR and STTR awards in nanotechnology and R&D related to cures for disease. Research priorities are to include projects relating to security, energy, transportation, and improving the security and quality of the water supply.
Oversight	An interagency SBIR/STTR Policy Committee is to be established by the Director of the Office of Science and Technology Policy. Funding is authorized for administrative, oversight, and contract processing costs. $27.5 million is authorized for FY2010 and for FY2011 for the SBA to transfer to the agencies for these activities. Oversight is required of small businesses that received multiple Phase I awards and no Phase II awards. Agency databases are to be created to support program evaluation and the Comptroller General is to audit agency calculation of extramural research.	The bill provides for the creation and maintenance of data bases collecting relevant information on the SBIR and STTR programs for use by both the government and the public sector. The Comptroller General is to audit agency calculation of extramural research.

Reauthorization Legislation

111th Congress Initiatives

The SBIR program has been extended several times and was scheduled to terminate on September 30, 2008. Although no specific legislation reauthorized the program, the Small Business Administration determined that P.L. 110-235 temporarily extended the SBIR activity through March 20, 2009.[3] Subsequently, P.L. 111-10 provided an additional extension of the program through July 31, 2009; P.L. 111-43 extended it through September 30, 2009; and P.L. 111-66 extended the effort through October 31, 2009. P.L. 111-89 once again extended the SBIR program through April 30, 2010, after which P.L. 111-214 extended the effort through September 30, 2010.

In the current Congress, H.R. 2965, the Enhancing Small Business Research and Innovation Act of 2009, passed the House, amended, on July 8, 2009. The Senate struck out all after the Enacting Clause and substituted the language of S. 1233 (the SBIR/STTR Reauthorization Act of 2009), amended, and passed H.R. 2965 in lieu on July 13, 2009. The following table provides a summary of selected provisions of these bills by issue.

110th Congress

In the 110th Congress, several bills were introduced to reauthorize and modify the SBIR initiative. H.R. 5819 passed the House on April 23, 2008, and S. 3362 was reported from the Committee on Small Business and Entrepreneurship on August 22, 2008. The following table is designed to highlight several of the differences and similarities between the two bills from the 110th Congress.

Issues for Consideration

Perhaps the most contentious issue in the reauthorization is that of the level of small business ownership by venture capital companies and eligibility under the SBIR and STTR programs. The original legislation establishing the SBIR program required that small firms must be at least 51% owned by an individual or individuals. Venture capital investment was permitted, and encouraged, but limited to 49% ownership interest. The intent of providing federal grant money, as explained in the House Science and Technology Committee report (H.Rept. 97-349, Part IV) to accompany the House bill that became the basis for the legislation, was

> to provide seed capital to small, high technology firms at the early, high risk stage of initial concept development. Funds provided ... would compensate for what has been described as a lack of investment capital for small businesses.... Funds provided under the SBIR program would be used ... to cover early development costs for small firms, providing early risk capital which is necessary for the procurement of follow-up support from the private sector.

Table 2. 110th Congress: Comparison of H.R. 5819 and S. 3362 by Selected Issue

Issue	H.R. 5819	S. 3362
Reauthorization	The termination date for the SBIR program is extended to September 30, 2010, while the STTR activity is extended to September 30, 2010.	The bill extends the SBIR program to September 30, 2022, while the STTR program is extended to September 30, 2023.
Award Levels	The level of awards made under the SBIR and STTR programs is increased from \$100,000 to \$300,000 for Phase I awards and from \$750,000 to \$2,200,000 for Phase II awards.	Phase I SBIR and STTR awards are increased from \$100,000 to \$150,000 and Phase II SBIR and STTR awards are increased from \$750,000 to \$1,000,000.
SBIR Set-Aside Amounts	No similar provision.	The set-aside is increased from 2.5% in 2009 by 0.1% per year up to 3.5% in 2019 with the following exceptions: the National Institutes of Health (NIH) remains at 2.5% and the increases in funding generated by the set-aside at the Department of Defense and the Department of Energy are not to be used for Phase I or II awards but for “activities that further the

Table 2. (Continued)

Issue	H.R. 5819	S. 3362
		technology readiness levels of technologies being developed under Phase II awards."
STTR Set-Aside Amounts	No similar provision.	The set-aside is increased from 0.3% to 0.4% for 2010 through 2011; to 0.5% for 2012 through 2013; and to 0.6% from 2014 on.
Sequential Awards	A recipient of a Phase I grant from one federal agency would be permitted to apply for a Phase II award from another agency to pursue the original work. Companies would be allowed to switch between the SBIR and STTR programs. A small firm may apply for a Phase II award without first obtaining and successfully completing a Phase I grant as currently required if certain conditions are met. Sequential Phase II awards for a project are permitted.	A recipient of a Phase I grant from one federal agency would be permitted to apply for a Phase II award from another agency to pursue the original work. A small business would be allowed to switch between the SBIR and STTR programs.
Majority Venture Capital Owned Small Businesses	For the SBIR and STTR programs, the bill would allow majority venture capital ownership in a small business if not more than 50% of the firm is owned by one venture capital company and the employees of the venture capital company are not a majority of the small firm's board of directors. If the venture capital company is controlled by a business with more than 500 employees, the small business is eligible only if not more than two large venture capital companies have ownership interest in the small firm, these large venture	The bill permits NIH to award not more than 18% of SBIR funds to majority venture capital-owned small businesses so long as "no single venture capital company owns more than 49 percent of the small business concern" upon a written determination provided to the Administrator of the SBA and the Senate Committee on Small Business and Entrepreneurship and the House Committee on Small Business. Allows other participating agencies to award not more than 8% of SBIR funds to majority venture capital-owned small

Table 2. (Continued)

Issue	H.R. 5819	S. 3362
	capital companies do not collectively own more than 20% of the small business, and the large venture capital companies "do not collaborate with each other to exercise more control over the small business concern than they could otherwise exercise individually."	businesses so long as "no single venture capital company owns more than 49 percent of the small business concern" upon a written determination provided to the Administrator of the SBA and the Senate Committee on Small Business and Entrepreneurship and the House Committee on Small Business.
Outreach	The bill authorizes and makes changes to the Federal and State Technology Partnership (FAST) program which provides grants to organizations to provide outreach designed to encourage increased participation in the SBIR program.	Similar provisions.
Commercialization Activities	To facilitate the commercialization of the results of the SBIR program, the bill requires agencies to establish procedures to encourage SBIR awardees to develop partnerships with other organizations, including prime contractors, business incubators, venture capital companies, and large business to assist them in moving to Phase III. "Express authority" is provided to agencies for development of "fast track" programs to end time delays between completion of Phase I and the award of Phase II grants. Agencies are required to implement a commercialization program to assist SBIR awardees in Phase III. A Minority Institution Pilot program is created to increase the number of SBIR and STTR	The bill provides for commercialization pilot programs for Phase II SBIR and STTR technologies in DOD and the civilian agencies. Encourages SBIR and STTR awards to small businesses that work with federal laboratories or are involved in cooperative research and development agreements (CRADAs).

Table 2. (Continued)

Issue	H.R. 5819	S. 3362
	applications from minority-owned small businesses.	
Areas of Concentration	Agencies are directed to focus on research in energy, rare diseases, transportation, and nanotechnology.	Agencies are encouraged to make SBIR and STTR awards in nanotechnology and R&D related to cures for disease.
Oversight	Agencies that administer SBIR grants of $50,000,000 or more must establish an SBIR Advisory Board comprised of agency employees, private sector representatives, veteran small business owners, and others to make recommendations on programmatic topics. An annual report is to be required.	The bill provides for the creation and maintenance of data bases collecting relevant information on the SBIR and STTR programs for use by both the government and the public sector.

However, the issue of venture capital ownership has come to the attention of Congress as the nature of venture capital investment and R&D performance has changed since the legislation was initially enacted in 1982. Recently, questions were raised as to whether small companies with majority ownership by venture capital companies met the eligibility requirements of the SBIR program. According to an April 2006 report by the Government Accountability Office (GAO-06- 5 65),

> in 2001, an SBA administrative law judge issued a decision clarifying that the terms "individuals" and "citizens" in the SBIR criteria meant only natural persons, not entities such as corporations.... Then in 2003, the same SBA administrative law judge issued a decision stating that venture capital firms could not be considered individuals for the purpose of satisfying the ownership criteria for the program.

In the 111th Congress, H.R. 2965, as passed by both the House and Senate, would alter the eligibility requirements regarding majority venture capital ownership of small firms in the SBIR and STTR program, although in different ways. Proponents of this change maintain that, particularly in the biotechnology sector, the most innovative companies are not able to use these programs because they do not meet these ownership criteria. They argue that because of the high cost of biotechnology R&D, large venture capital investments are often a necessity for many of these firms. By excluding such companies from the SBIR and STTR programs, advocates maintain that the pool of applicants for participation in the effort, specifically at the National Institutes of Health, is decreasing with detrimental effects to the health and strength of the U.S. economy.

Opponents of altering the eligibility requirements argue that the program is designed to provide financial assistance where venture capital is not available. These experts assert that the program's objective is to bring new concepts to the point where private sector investment is feasible. However, they claim that the changes in the bill associated with venture capital majority ownership take federal support away from actual small businesses and place government funds into the hands of investors while favoring the biotechnology and venture capital sectors.

The Senate-passed version of H.R. 2965 would increase the set-aside amounts for both the SBIR and STTR programs while both versions increase the amount of individual SBIR and STTR Phase I and Phase II awards that may be made. Raising the upward limit of the permitted awards appears to reflect a recognition that the performance of research and development has become increasingly expensive over the 25 years since passage of the original legislation. However, this may result in fewer awards available to small firms through these two programs, particularly under H.R. 2965 as passed by the House, which does not increase the amount of the set-aside.

Both bills provide more congressional direction to the participating agencies as to topics for inclusion in program solicitations and to the characteristics to be considered during the selection process. Outreach to specific states, rural areas, and identified groups is mandated. Increased commercialization activities would be fostered by provisions in both versions of H.R. 2965, and additional mechanisms and/or funding for oversight are included.

As the 111th Congress considers the reauthorization of the SBIR and STTR programs, and any possible changes in program operation, these and other issues may continue to be explored.

End Notes

[1] For a more detailed discussion of the current SBIR program see CRS Report 96-402, *Small Business Innovation Research (SBIR) Program*, by Wendy H. Schacht.

[2] It should be noted that P.L. 97-2 19 excluded appropriated funds for defense programs in the Department of Energy from that agency's extramural R&D calculations.

[3] Letter from Peter K. Schalestock, Acting General Counsel, U.S. Small Business Administration, September 11, 2008, http://grants.nih.gov/grants/funding

In: U.S. Innovation Programs
Editors: Allison E. Berger et al.
ISBN: 978-1-61122-809-0

Chapter 5

INTELLECTUAL PROPERTY IN INDUSTRIAL DESIGNS: ISSUES IN INNOVATION AND COMPETITION

John R. Thomas

SUMMARY

Under current intellectual property laws, industrial designs may potentially be protected through design patents, trade dress, and copyright. In addition, the Vessel Hull Design Protection Act established a specialized, or *sui generis*, intellectual property right for the protection of boat hull designs. Some experts argue that the present intellectual property regime does not adequately protect industrial designers.

In the 111th Congress, two bills would establish a "repair" exemption within the Patent Act. Although potentially of broad application, H.R. 3059 and S. 1368, each titled the "Access to Repair Parts Act," appear to have been motivated by the enforcement of design patents that are said to restrict competition in the secondary market for automobile replacement parts.

Also in the 111th Congress, legislation has been introduced that would establish proprietary rights in fashion designs. H.R. 2196, the Design Piracy Prohibition Act, would provide a three-year term of protection for fashion designs based upon registration with the U.S. Copyright Office.

The Senate has also recently provided its consent to ratify the Geneva Act to the Hague Agreement Concerning the International Registration of Industrial Designs, an international agreement that would allow U.S. designers more readily to obtain intellectual property protection overseas. Accession to this international agreement would require changes to the design patent statute, including expansion of the term of design patents from 14 to 15 years.

Recent, significant judicial developments concerning the scope of protection accorded to design patents have also occurred.

Supporters of *sui generis* regimes assert that they allow for protection of subject matter, like industrial designs, that at times fall outside traditional intellectual property paradigms. Proponents believe that these systems can prevent levels of free riding that ultimately discourage innovation. However, some observers are concerned that the proliferation of such systems may limit the ability of others to compete, ultimately diminishing consumer choice.

In recent years, Congress has expressed interest in intellectual property protection for industrial designs. In the 111th Congress, legislation has been introduced that would establish a "repair" exemption within the Patent Act.[1] Although potentially of broad application, this legislation appears to have been motivated by intellectual property rights in the designs of automobile parts.[2] Also in the 111th Congress, proposed legislation would provide proprietary rights in fashion designs.[3] The Senate has also recently provided its consent to ratify the Geneva Act to the Hague Agreement Concerning the International Registration of Industrial Designs,[4] an international agreement that would allow U.S. designers to obtain intellectual property protection overseas more readily.[5] In addition to these legislative activities, industrial designs have also been subject to notable developments in the federal courts and internationally. A review of intellectual property issues pertaining to industrial designs therefore appears timely.

The term "industrial design" broadly refers to the creation of the form and function of objects of everyday use.[6] Virtually every manufactured product, including such items as automobiles, clothing, computers, footwear, furniture, and telephones, embodies an industrial design.[7] Industrial designers perform the tasks previously accomplished by individual artisans with respect to mass-produced goods, commonly endeavoring to make products more convenient to use and more visually appealing.[8] As explained by the Industrial Designers Society of America:

> Industrial design is the profession that determines the form of a manufactured product, shaping it to fit the people who use it and the

> industrial processes that produce it. Industrial Designers work to make our lives more comfortable, pleasurable and efficient. By studying people at work, at home and in motion, they create products like office chairs that promote proper posture, kitchen tools that are comfortable even for elderly hands and toys that provide safe play and learning for all children. In particular, Industrial Designers deal with the parts of a product that humans interact with, striving to give universal access to products that are ecologically responsible and safe to use. Also, they give a product with distinctive elegance that makes us want it.[9]

"Intellectual property" is a phrase that refers to proprietary rights in creations of the mind.[10] Currently, some measure of protection for industrial designs can be found under three different intellectual property laws: copyright, trademark, and patent.[11] Perceived shortcomings with these systems as applied to industrial designs have led to proposals to establish specialized, or *sui generis*,[12] regimes that apply to particular sorts of designs. Congress established such a regime for boat hulls in 1988,[13] and serious discussion is ongoing with respect to the merits of *sui generis* intellectual property rights for fashion designs.[14]

This chapter identifies several current issues relating to intellectual property in industrial designs. It begins by describing the different sorts of intellectual property protection that apply to industrial designs. The report then identifies current issues at the interface between intellectual property and industrial designs. The report closes by reviewing the impact that *sui generis* rights regimes may have upon innovation and competition in the United States.

Industrial Designs and Intellectual Property

Industrial designs that are well-received in the marketplace may attract competition. William T. Fryer III, a member of the faculty of the University of Baltimore School of Law, asserts that "[o]nce an effective product design is successful, competitors rush in and copy the product, particularly the product appearance."[15] Advocates of enhanced protection for industrial designs therefore assert that intellectual property rights potentially provide an important mechanism for allowing innovators to reap the economic rewards of their investments.[16] On the other hand, some observers are concerned that overly expansive intellectual property rights may raise prices and decrease the availability of consumer goods.[17]

Under current intellectual property laws, industrial designs may potentially be protected through copyright, trade dress, and design patents.[18] Congress has also established a specialized, or *sui generis*, intellectual property right for the protection of boat hull designs.[19]

COPYRIGHT

The Copyright Act of 1976 provides protection for original works of authorship. The types of creations addressed by copyright range from traditional works of art, including literature, music and visual art, to such modern forms of artistic expression as sound recordings, motion pictures and computer software.[20] Copyright protection arises automatically, as soon as the work has been fixed in tangible form.[21] Authors may register their works with the U.S. Copyright Office, however, and obtain certain procedural and substantive advantages during copyright enforcement.[22] The copyright law affords authors the exclusive right to reproduce, adapt, and publicly distribute, perform, and display the protected work, subject to limitations such as the fair use privilege.[23] The term of copyright is ordinarily the life of the author plus 70 years.[24]

The availability of copyright protection for industrial designs has been described as limited, however.[25] The Copyright Act of 1976 provides protection for "pictorial, graphic, and sculptural works" but further explains that:

> the design of a useful article ... shall be considered a pictorial, graphic, or sculptural work only if, and only to the extent that, such design incorporates pictorial, graphic, or sculptural features that can be identified separately from, and are capable of existing independently of, the utilitarian aspects of the article.[26]

The Court of Appeals for the Ninth Circuit has explained that "[t]his limitation is in keeping with the notion that functional items are not eligible for the relatively long-term protections of copyright, as opposed to the more temporary rights provided by the Patent Act."[27] Because industrial designs are commonly integrated into a "useful article"[28]—such as automobiles, handheld electronic devices, or clothing—they are often not eligible for copyright.[29]

Trade Dress

Protection for industrial designs may also occur under the trademark law in the form of "trade dress" protection. The term "trade dress" refers to the overall visual image that a product presents to consumers, including its design.[30] To be entitled to trademark protection, a significant number of consumers must perceive the trade dress as denoting the product of a particular manufacturer, rather than being arbitrary or merely decorative, after the trade dress has been used in the market.[31] In trademark law, this notion of acquired distinctiveness is termed "secondary meaning."[32]

Trademark rights in trade dress may arise under state law as soon as the product is used in commerce. However, trade dress may be registered with the United States Patent and Trademark Office (USPTO), a step that affords significant substantive and procedural advantages.[33] The trade dress owner may prevent others from using a similar product appearance that creates a likelihood of confusion as to the source or sponsorship of the associated goods.[34] These rights persist so long as the trade dress continues to be used and retains its distinctiveness.[35]

Commentators have identified a primary limitation of trade dress protection with respect to industrial designs.[36] Trade dress rights cannot be established in functional features of a product. Courts have reasoned that such protection would overly hamper competition and undermine the policies of the patent system.[37] As a result, if the industrial design makes the product easier to manufacture, more convenient to ship, or work better, then it may not be subject to trade dress protection.[38]

Design Patents

In addition to granting so-called "utility patents" directed towards machines, manufactures, compositions of matter and processes,[39] the Patent Act of 1952 also allows for "design patents." An inventor may obtain a design patent by filing an application with the USPTO directed towards a "new, original and ornamental design for an article of manufacture."[40] Most design patent applications consist primarily of drawings that depict the shape or surface decoration of a particular product. They may concern any number of products, including "apparel, automobile parts, computer products, containers, cosmetics, electronics products, textile designs, home furnishings, home appliances, jewelry, motor vehicles, office supplies, optics and toys."[41]

To obtain protection, the design must not have been obvious to a designer of ordinary skill of that type of product.[42] In addition, a design must be "primarily ornamental" to be awarded design patent protection.[43] Similar to the doctrine of functionality in trademark law, if the design is instead "dictated by the performance of the article, then it is judged to be functional and ineligible for design patent protection."[44] In such circumstances the designer should pursue a utility patent, which protects the functional and utilitarian aspects of products and processes.[45]

Issued design patents confer the right to exclude others from the "unauthorized manufacture, use, or sale of the article embodying the patented design or any colorable imitation thereof."[46] The scope of protection of a design patent is provided by drawings within the design patent instrument.[47] To establish infringement, the design patent proprietor must prove that "in the eye of an ordinary observer, giving such attention as a purchaser usually gives," the patented and accused designs "are substantially the same, if the resemblance is such as to deceive such an observer, inducing him to purchase one supposing it is the other."[48] The courts also require that "no matter how similar two items look, the accused device must appropriate the novelty in the patented device" that distinguishes it from earlier designs.[49] The term of a design patent is 14 years from the date the USPTO issues the patent.[50]

Intellectual property attorneys Richard Sybert and Lindsay Hulley have explained that the "best protection for product design is likely a design patent in the United States...."[51] Design patents have been described as less expensive to prepare than utility patents, and unlike utility patents they are not subject to "maintenance fees" that must be submitted to the USPTO during the life of the patent.[52] In contrast to copyright enforcement, design patent infringement does not require a showing that the defendant accessed the prohibited design and copied it.[53] And unlike trademark, the design patent holder need show neither the existence of consumer confusion nor the design's acquisition of "secondary meaning" in order to prove infringement.[54]

The design patent regime has nonetheless been subject to criticism by industry representatives, lawyers, and academics alike. Some commentators have expressed concerns that the standards of novelty and nonobviousness, which the design patent law essentially adapted from the law of utility patents, do not provide apt or fair evaluating criteria for industrial designs. As Gerard Magliocca, a member of the faculty of the Indiana University School of Law—Indianapolis, explains:

> To see this point more clearly, think of these two elements as part of a broader test to determine whether an improvement represents an "advance" over the state of the art in a given field. Viewed in these terms, one commentator has observed that, "[w]hereas it may often be possible to recognize a technological innovation as an 'advance' because of its efficiency in promoting economies or in achieving previously unattainable utilitarian ends, there exist no comparable criteria of 'advance' for artistic creations." Indeed, most designs simply involve rearranging basic artistic elements (*e.g.*, colors, shapes, and materials) into a new pattern. The result of this reshuffling process can almost always be considered obvious—and hence unpatentable—because any designer could have come up with the combination at issue. Put another way, the nonobviousness and novelty requirements often end up restricting design protection to those rare instances when an improvement establishes an entirely new style. This leaves most designers out of luck and is at odds with the reality that commercial art is generally incremental in nature.[55]

Other observers believe that acquiring design patents remains an overly costly and time- consuming procedure. These factors may reduce the attractiveness of such protection for products, like clothing, that remain popular for only short periods of time. [56] Design patent law's functionality doctrine, which is similar to principles found in copyright and trademark law, has again been subject to critique. An industrial design that successfully integrates functional and ornamental aspects may simply be ineligible for protection.[57]

Vessel Hull Design Protection

In the boating industry, some manufacturers have been known to use a "plug molding" technique to duplicate the hulls of their competitors. Some observers believe that the "boat manufacturing industry has long been plagued by low-cost boat makers who think nothing of taking a successful competing boat hull design to make a direct-cast mold for their own unauthorized manufacturing use."[58] In response, the state legislature of Florida enacted legislation that prohibited certain uses of the plug molding process.[59] However, the 1989 Supreme Court ruling in *Bonito Boats Inc. v. Thunder Craft Boats, Inc.* held that this legislation was preempted by the federal patent statute and therefore invalid.[60]

Following the *Bonito Boats* decision, Congress enacted the Vessel Hull Design Protection Act.[61] This 1998 statute, currently codified in Title 13 of the Copyright Act, allows for the registration of an original design that has actually been incorporated into a fabricated vessel hull.[62] Although this system is administered by the Register of Copyrights, the protection afforded by the Vessel Hull Design Protection Act is not a copyright, but rather a *sui generis* right.[63]

In order to enjoy protection, the design owner must affix notice of the registration upon vessel hulls that embody the design.[64] The protected design affords its owner the exclusive right to "make, have made, import, for sale or for use in trade, any useful article embodying that design" and "sell or distribute for sale or for use in trade any useful article embodying that design."[65] The statute does not extend protection to designs that are "staple or commonplace" or "dictated solely by a utilitarian function."[66] In addition, an accused design that is "original and not substantially similar in appearance to a protected design" is exempted from infringement.[67]

The Vessel Hull Design Act also creates infringement liability if a seller or distributor who did not itself make or import an infringing article induces, or acts in collusion with, a manufacturer to make or import the article.[68] A seller or distributor may also be liable for infringement if it fails to respond to the design owner's request to disclose the source of an infringing article, and if that person orders or reorders the article after receiving notice of the protected design.[69] The act also establishes a limited "innocent infringement" exemption which excuses individuals who made, imported, sold, or distributed an article embodying the copied design without knowledge that the design was protected.[70]

The design owner may bring a lawsuit in federal court against those it believes are infringing.[71] Remedies available under the act include an injunction[72] and damages adequate to "compensate for the infringement."[73] The act also allows the court to increase the damages "to such amount, not exceeding $50,000 or $1 per copy, whichever is greater, as the court deems to be just," provided that the damages awarded are compensatory rather than punitive in nature.[74] The term of protection is ten years.[75]

Some experts believe that in the decade since its enactment, the Vessel Hull Design Protection Act had been "underutilized, no doubt in part due to the difficulty in proving infringement based on the way in which 'hull' was originally defined."[76] In particular, the act currently defines the term "hull" as "the frame or body of a vessel, including the deck of vessel...."[77] This definition required that courts consider the features incorporated into a vessel's deck when considering what subject matter is protected. [78] As a result, the act has been said to have created a "legal loop-hole for 'knock-off' artists to freely copy a protected boat hull design as long as they are clever enough to modify one or more protectable elements of the 'deck.'"[79]

P.L. 110-434, the Vessel Hull Design Protection Act Amendments of 2008, apparently responded to these concerns. Enacted by the 110th Congress, that legislation clarified the distinction between a hull and a deck. [80] P.L. 110-434 also specified that protection is available for designs of hulls and decks, either individually or in combination. President Bush signed into law the amendments to the Vessel Hull Design Protection Act on October 16, 2008.[81]

Current Issues in Industrial Design Protection

Multiple forms of intellectual property protection are potentially available to industrial designs. Some observers believe that this "ragged quilt of protection" does not adequately protect designers, however.[82] In particular, the inability of copyright, trademark, and design patent law to protect industrial designs that have some functional component is frequently criticized.[83] Others disagree, believing that there is no compelling case for expanding intellectual property protection for industrial designs and that doing so would limit competition and the availability of products for consumers.[84] Recent debate about the scope of intellectual property rights for industrial designs has occurred within several specific contexts. This chapter next turns to a review of these issues.

Intellectual Property Rights in Fashion Designs

Congress has recently considered legislation that would, for the first time in the United States, provide explicit intellectual property protection for fashion designs.[85] In the 111th Congress, H.R. 2196, titled the Design Piracy Prohibition Act, would have expanded the scope of coverage under the Vessel Hull Design Protection Act to also include "an article of apparel."[86] This term is broadly defined to include an article of clothing, including undergarments, outerwear, gloves, footwear, headgear, handbags, purses, tote bags, belts, and eyeglass frames.[87] In order to secure protection, the designer must apply for registration within six months after the article has been made available to the public.[88] The term of protection is set at three years.[89]

Perceived shortfalls in the established intellectual property regimes provided the impetus for this Design Piracy Prohibition Act.[90] Design patents have been described as "difficult and expensive to obtain," as well as entailing "a lengthy examination process."[91] Trade dress law may protect a fashion design if that design has obtained a reputation among consumers as identifying the source of the product.[92] However, such protection is not of general applicability, and its extent is limited to uses of the design that create a substantial likelihood of consumer confusion.[93] Finally, the prohibition against obtaining copyright for "useful articles" means that clothing most often does not secure protection.[94]

Due to the limited nature of existing intellectual property rights for fashion, many designers assert that their successful designs are "knocked off" in a matter of weeks or even days.[95] This trend may particularly impact young designers who have yet to establish reputations in the marketplace, and in fact may be unable to do so when competitors frequently copy their work.[96] Proponents of the Design Piracy Prohibition Act further asserted that the relatively short term of protection afforded—three years—will protect high end "haute couture" designs but not unduly impact the public.[97] Supporters of this statute also point to intellectual property rights that have been established for fashion design overseas.[98]

Not everyone agrees that intellectual property rights of the sort contemplated by the Design Piracy Protection Act would be appropriate. Opponents of the legislation explained that the fashion industry is enjoying great success and that copying is an integral and accepted practice within it.[99] They also claim that this legislation would have led to numerous accusations of infringement, stifling the creative production of fashion designs and reducing consumer choice.[100] They further observe that substantial copying of designs continue to occur in foreign states that have enacted fashion design protection laws, suggesting that this intellectual property right may prove of limited effectiveness.[101]

U.S. Adherence to the Hague Convention

An international agreement known as the Geneva Act to the Hague Agreement Concerning the International Registration of Industrial Designs potentially simplifies the process through which industrial design protection can be obtained in many countries.[102] Under current law, U.S. designers must file separate applications for intellectual property protection in each jurisdiction where they seek to obtain rights.[103] This process is potentially complex and may involve multiple languages, currencies, and local legal representatives.[104]

The Geneva Act to the Hague Convention would instead provide U.S. designers with the option of filing a single English-language application at the U.S. Patent and Trademark Office (USPTO). The application would then be forwarded to the World Intellectual Property Organization (WIPO), which then publishes the application. The intellectual property authorities of each designated member state then have a maximum of 12 months to provide notice that the application has been refused. In the event of a refusal, further efforts to obtain intellectual property protection in that design would occur in keeping with national laws. [105] If a member state does not give notice of a refusal within 12 months, then the application is deemed to be granted.[106]

The Senate gave advice and consent to the ratification of the Geneva Act to the Hague Convention on December 7, 2007.[107] The United States has not yet deposited its instrument of ratification with WIPO, however. As explained by President Bush's letter of transmittal to the Senate:

> In the event that the Senate provides its consent to ratify the Agreement, the United States would not deposit its instrument of ratification until the necessary implementing legal structure has been established domestically.[108]

The USPTO has testified before Congress that a number of changes would need to be made to render U.S. design patent law compatible with the Geneva Act to the Hague Convention, including increasing the term of protection from 14 to 15 years and providing limited rights to patent applicants between the date their international applications are published and the date the patent issued.[109] This implementing legislation has not yet been introduced before Congress.

Automobile Spare Parts

The use of design patents to protect automobile aftermarket or collision repair parts has long been debated.[110] Some observers believe that allowing intellectual property rights to cover "must- match" spare parts on complex products overly limits competition. Others believe that distinguishing between primary and secondary markets for the protected design would be unprecedented and unnecessarily limit protection for designers.[111] This debate has proceeded overseas as well. On December 12, 2007, the European Parliament approved a proposal to end design protection for automobile spare parts and other machinery components.[112]

Recent legal controversies have renewed this debate domestically. In *In re Certain Automotive Parts*,[113] the International Trade Commission held that certain third-party replacement automobile parts, designed for use in the 2004 Ford F-150 pickup truck, infringed design patents owned by the Ford Motor Co. As a result, the ITC banned the importation of these component parts into the United States. In response, a coalition of automobile part manufacturers called on Congress to add a "repair" infringement exemption to U.S. design patent law.[114]

In the 111th Congress, legislation was introduced that would have introduced a new subsection (j) to the "Infringement of patent" provision at § 271 of the Patent Act.[115] That subsection would read:

> It shall not be an act of infringement to make, use, offer to sell, or sell within the United States or import into the United States any article of manufacture that itself constitutes a component part of another article of manufacture, if the sole purpose of the component part is for the repair of the article of manufacture of which it is a part so as to restore its original appearance.

It should be appreciated that this exemption appears to extend beyond automobile spare parts and design patents. Rather, proposed § 271(j) of the Patent Act would apparently have applied to all industries with respect to any patent.[116]

Observers vary as to the merits of this proposal. Supporters of the legislation assert that design patents on automobile replacement parts "increase repair bills, block competition, and threaten the safety of consumers who may forego necessary repairs because of their high cost."[117] However, not all concerned parties agree. Design patent attorney Perry Saidman reportedly has stated that the legislation would ignore "the rationale behind design patents" and opined that "any company whose business is predicated on copying the original designs of others will ... beat a path to Congress' door, claiming that their knockoff products increase competition and reduce prices for consumers."[118]

Judicial Developments Concerning Design Patents

Some legal experts have drawn attention to recent judicial opinions with respect to design patents.[119] One of these opinions, *Arminak and Associates, Inc. v. Saint-Gobain Calmar, Inc.*,[120] concerned the identity of the "ordinary observer" in design patent infringement determinations. In order to decide that an accused design infringes a design patent, a court must conclude that "in the eye of an ordinary observer, giving such attention as a purchaser usually gives," the patented and accused designs "are substantially the same, if the resemblance is such as to deceive such an observer, inducing him to purchase one supposing it is the other."[121] The Supreme Court has held that "ordinary observers" should not be viewed as experts versed in the designs used within a particular industry, but rather "those who buy and use" the product that bears the particular design. [122]

In the *Arminak* decision, the Court of Appeals for the Federal Circuit ("Federal Circuit") held that under the facts of that case, an industrial purchaser was the appropriate "ordinary observer" for assessing infringement. The design patent in *Arminak* depicted the top portion of a trigger sprayer, which is commonly termed a "shroud." Both the patent proprietor and the accused infringer sold trigger sprayers to producers of liquid household products. The Federal Circuit concluded that ordinary consumers, who do not purchase shrouds by themselves, but rather household cleaning products with trigger sprayers, could *not* be judged the "ordinary observer." Rather, "the purchaser of the patented and accused designs in this case is the purchaser of one of a retail product's component parts that is thereafter assembled with other parts to make the retail product."[123] The Federal Circuit therefore affirmed the district court's finding of noninfringement.

The ruling in *Arminak* has been criticized due its potential impact upon design patent holders. As industrial purchasers are considerably more sophisticated than ordinary consumers concerning the nuances of particular product designs, a finding of infringement is much less likely when the standard of the "ordinary observer" is raised beyond the level of the lay public. Although the *Arminak* holding may potentially be justified under its particular facts, design patent attorney Perry Saidman has viewed this ruling as inconsistent with governing Supreme Court cases.[124] Mr. Saidman has further concluded that the *A rminak* holding effectively renders "a holding of design patent infringement virtually unattainable, ... indeed destroying all the protection for designs that Congress intended to give."[125]

Another judicial opinion, *Egyptian Goddess, Inc. v. Swisa, Inc.*,[126] raised two issues of current controversy with respect to design patents. In *Egyptian Goddess* and other decisions, courts have engaged in a process of "claim construction" with respect to design patents.[127] Because the claims of design patents consist of visual images of the protected design, this "claim construction" process consists of putting into words the visual appearance of the designs shown in the patent's drawings. This textual description is then compared to the accused product under the "ordinary observer" standard in order to determine whether the design patent was infringed or not. For example, the asserted design patent in the *Egyptian Goddess* case concerned an "ornamental nail buffer." One of the patent's seven figures appeared as follows:

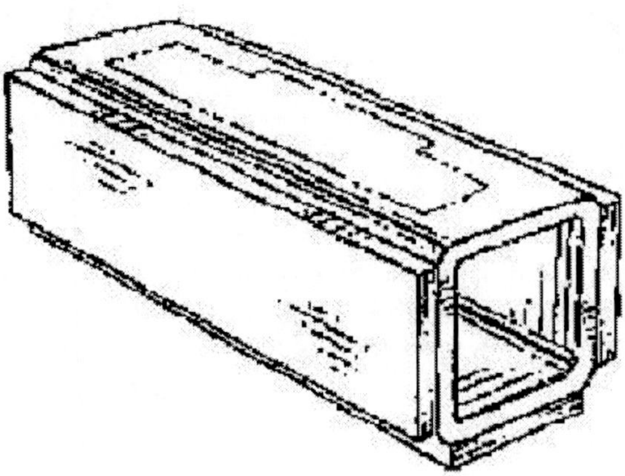

The claim construction associated with the design patent was as follows:

> A hollow tubular frame of generally square cross section, where the square has sides of length S, the frame has a length of approximately 3S, and the frame has a thickness of approximately T = 0.1 S; the corners of the cross section are rounded, with the outer corner of the cross section rounded on a 90 degree radius of approximately 1 .25T, and the inner corner of the cross section rounded on a 90 degree radius of approximately 0.25T; and with rectangular abrasive pads of thickness T affixed to three of the sides of the frame, covering the flat portion of the sides while leaving the curved radius uncovered, with the fourth side of the frame bare.[128]

Some observers believe that the judicial claim construction inquiry potentially provides a more fair and predictable method for determining the scope of patent protection.[129] However, others have asserted that this "verbalization" of the claimed design overly limits the scope of design patents and does not adequately allow for the judgment of the "ordinary observer" in design patent infringement determinations.[130]

In *Egyptian Goddess*, the Federal Circuit concluded that trial courts possessed discretion on whether to conduct a detailed claim construction or not. Judge Bryson explained that "[g]iven the recognized difficulties entailed in trying to describe a design in words, the preferable course ordinarily will be for a district court not to attempt to 'construe' a design patent claim by providing a detailed verbal description of the claimed design."[131] However, "a district court's decision regarding the level of detail to be used in describing the claimed design is a matter within the court's discretion.... "[132]

The *Egyptian Goddess* opinion also changed the way that courts will conduct infringement analyses of design patents in the future. Previous, the Federal Circuit had used the "point of novelty" standard when deciding whether design patents were infringed or not.[133] Under this standard, "no matter how similar two items look, the accused device must appropriate the novelty in the patented device" that distinguishes it from earlier designs in order to prove infringement.[134] The Federal Circuit had explained that the patent proprietor must set forth either a single novel design element or a combination of known elements that collectively present a "non-trivial advance over the prior art."[135] If the design element or elements asserted by the patent proprietor do not meet this standard, then the court will reach a judgment of noninfringement.

In the *Egyptian Goddess* case, the patent proprietor's asserted point of novelty was the combination of four of the claimed design's elements: (1) an open and hollow body, (2) square cross-section, (3) raised rectangular pads, and (4) exposed corners. Applying Federal Circuit precedent, the trial court determined that an earlier design patent directed towards a nail buffer disclosed all of these elements, with the exception that it was triangular, rather than square in cross-section. Observing that a number of other references illustrated nail buffers with square cross-sections, the court rejected the asserted "point of novelty" and rendered a judgment of noninfringement.[136]

As it had come to be applied by the Federal Circuit, the "point of novelty" standard had been criticized on a number of grounds. Some commentators believed that this approach confused validity and infringement standards for design patents and also resulted in a narrow scope of design patent protection.[137] However, other observers explained that design patents contain general illustrations that do not specify the distinctions between the patented design and the state of the art. They favored a rule that ensures that there will be no finding of infringement when the accused design has the same overall appearance of both the design patent and earlier designs, particularly because they both use features found in well-known predecessors.[138]

In response to critiques of the "point of novelty" standard, the Federal Circuit explicitly rejected this test in *Egyptian Goddess*. The Court of Appeals instead concluded that "the 'ordinary observer' test should be the sole test for determining whether a design patent has been infringed."[139] Although some commentators believe this ruling is not of particular practical significance,[140] other observers believe that *Egyptian Goddess* had made design patents more attractive because, in their view, it is now easier for a court to reach a conclusion of infringement.[141]

ISSUES IN INNOVATION AND COMPETITION

Current congressional interest in industrial designs spans a number of different issues and industries. This breadth of interest in strengthening intellectual property rights for industrial designs arguably reflects the changing nature of the U.S. economy. Given the increasing sophistication of foreign manufacturing, the protection of industrial designs provides one mechanism for sustaining competitive domestic design-based industries. As attorney Regan Keebaugh explains, "[i]f economic reality means the United States can no longer compete in the production side of manufacturing, then one way to maintain jobs in the manufacturing industry is through developing a strong design industry to ensure that products made overseas are designed in the United States."[142]

Intellectual property laws are said to promote innovative industries by establishing exclusive rights in new inventions, industrial designs, and other creations. Absent such rights, "free riders" could easily duplicate and exploit the innovative developments of others. Further, because they incurred no design and development costs, copyists could likely undersell the original innovator. The resulting inability of innovators to capitalize on their creations would lead to an environment where too little research and development occurs.[143] Intellectual property rights may also encourage others to design around these proprietary interests, pointing the way to new products, markets, economies of production and even entire industries.[144]

Although the established intellectual property paradigms of copyrights, trademarks, and patents provide protection for traditional works of authorship, brand names, and inventions, they also limit or deny protection for other sorts of subject matter. As this chapter has explained, copyright, trade dress, and design patent law to varying degrees deny protection to industrial designs with a functional component. [145] *Sui generis* protection systems may potentially fill gaps that may not have been anticipated by the framers of the traditional intellectual property regimes.[146]

Sui generis protection regimes also offer a potential advantage when compared to the copyright, trademark, and patent statutes. Although the established forms of intellectual property apply broadly to numerous industries and types of innovations,[147] *sui generis* protection regimes may be more precisely tailored to meet the needs of particular industries. For example, the acquisition of rights could be designed to fit the perceived needs of particular industries, the scope and term of protection may be particular product cycles, and competitors and consumers could enjoy the privilege to use protected subject matter in certain circumstances.[148]

Policy makers may also appreciate that the designs of previous *sui generis* regimes have tended to be based upon one of the established intellectual property paradigms. For example, the rights established under the Semiconductor Chip Protection Act of 1984[149] are similar to copyright. [150] On the other hand, both the design and plant patent systems were incorporated into the existing utility patent laws and therefore follow a patent-like model.[151] Some observers believe that copyright- or trademark-like protection represents the most appropriate scheme of protection for industrial designs because neither regime requires courts or government officials to analyze the merits of "creative expression applied to goods."[152] On the other hand, other experts believe that the design patent law, which does entail an analysis of whether an industrial design would have been obvious to a skilled artisan,[153] has historically served as the "best option" for protecting industrial designs.[154]

Although the establishment of *sui generis* regimes may reduce harmful free riding that results in reduced investment in innovation, experts have also expressed concerns about the expansion of intellectual property rights. They observe that the primary mechanism through which intellectual property rights create incentives to innovate is to limit the ability of others to participate in the marketplace.[155] From the perspective of the government, such systems are inexpensive when compared to prizes or other potential incentives.[156] Yet they also limit the ability of firms to compete and possibly raise prices and limit the products available to the public, at least in the short term.[157]

In this regard, it should be appreciated that the traditional intellectual property regimes often did not leave certain subject matter outside the scope of protection by accident. Rather, this circumstance resulted from a purposeful legislative choice. [158] For example, the patent statute stipulates that designs that would have been obvious to a skilled artisan should not be awarded proprietary rights.[159] The Supreme Court recently explained this rule as resulting from the judgment that "[w]ere it otherwise patents might stifle, rather than promote, the progress of useful arts."[160] To the extent that proposed *sui generis* regimes lower this threshold and award proprietary rights in "subpatentable" innovations, they may diminish the public domain and hamper both future innovation and competition.[161]

Finally, congressional award of proprietary rights to some industries may inspire additional lobbying efforts by others. Other sorts of potentially innovative products—including databases and other compilations of information;[162] perfumes, colognes, and scents;[163] and computer software[164]—are among those that, at least according to some observers, do not interface well with the traditional copyright, trademark, and patent regimes. Mark Janis, a member of the faculty of the University of Iowa College of Law, explains that once one industry has made its case for a *sui generis* regime of intellectual property rights, "[i]t takes little imagination to extend this reasoning to justify the creation of a multitude of additional second tier ... regimes having specialized subject matter requirements."[165] Determining which industries, if any, deserve *sui generis* rights outside the established intellectual property paradigms ultimately remains a matter of legislative judgment.

End Notes

[1] H.R. 3059; S. 1368 (both titled the "Access to Repair Parts Act").

[2] *See* "Legislation/Patents: Auto Parts Groups Lobby for Bill to Exempt Repair Parts from Patent Infringement," 75 *Patent, Trademark & Copyright Journal* (April 25, 2008), 674.

[3] H.R. 2196 (titled the "Design Piracy Prohibition Act").

[4] World Intellectual Property Organization, Hague Agreement Concerning the International Deposit of Industrial Designs, July 2, 1999, 2279 U.N.T.S. 156 (available at http://www.wipo.int/export/sites/www/hague/en/legal_texts/pdf/ geneva_act_1999.pdf).

[5] Treaty No. 109-2 1 (December 7, 2007).

[6] *See, e.g.,* Henry Dreyfuss, *Designing for People* (Allworth Press 2003); Carma Gorman, ed., *The Industrial Design Reader* (Allworth Press 2003).

[7] Regan E. Keebaugh, "Intellectual Property and the Protection of Industrial Design: Are Sui Generis Protection Measures the Answer to Vocal Opponents and a Reluctant Congress?," 13 *Journal of Intellectual Property Law* (2005), 258.

[8] William T. Fryer, "The Hague Agreement on the Protection of Industrial Designs: Strategies to Use and U.S. Choices in Ratification of the Geneva Act," 89 *Journal of the Patent and Trademark Office Society* (2007), 662.

[9] Industrial Design Society of America, "What is ID?" (available at *http://www.idsa.or g/webmodules/articles/* articlefiles/what_is_id_brochure.pdf).

[10] *See* Roger E. Schechter and John R. Thomas, *Intellectual Property: The Law of Copyrights, Patents and Trademarks* (Thomson West 2003), 1.

[11] Keebaugh, *supra*, at 260.

[12] The term *sui generis*, which literally means "of its own kind," "is used in intellectual property law to describe a regime designed to protect rights that fall outside the traditional copyright, trademark, patent, and trade secret doctrines." *Black's Law Dictionary* (8th ed. 2004).

[13] The Vessel Hull Design Protection Act was enacted as Title V of the Digital Millennium Copyright Act of 1998 in the 105th Congress. P.L. 304, 112 Stat. 2905 (1998).

[14] *See* Lynsey Blackmon, "The Devil Wears Prado: A Look at the Design Piracy Prohibition Act and the Extension of Copyright Protection to the World of Fashion," 35 *Pepperdine Law Review* (2007), 107.

[15] Fryer, *supra*, at 662.

[16] Perry J. Saidman, "The Crisis in the Law of Designs," 89 *Journal of the Patent and Trademark Office Society* (2007), 301; David Goldenberg, "The Long and Winding Road: A History of the Fight Over Industrial Design Protection in the United States," 45 *Journal of the Copyright Office Society* (1997), 21.

[17] Keebaugh, *supra* at 256 (noting those views).

[18] *See* Orit Fischman Afori, "Reconceptualizing Property in Designs," 25 *Cardozo Arts & Entertainment Law Journal* (2008), 1105.

[19] *See* Bradley J. Olson, "The Amendments to the Vessel Boat Hull Protection Act of 1988: A New Tool for the Boating Industry," 38 *Journal of Maritime Law and Commerce* (2007), 177.

[20] 17 U.S.C. § 102 (2006).

[21] *Id.*

[22] 17 U.S.C. §§ 408—412 (2006).

[23] 17 U.S.C. §§ 106, 107—122 (2006).

[24] 17 U.S.C. § 302 (2006).

[25] Emily S. Day, "Double-Edged Scissor: Legal Protection for Fashion Design," 86 *North Carolina Law Review* (2007), 245-47; Keebaugh, *supra*, at 264-65.

[26] 17 U.S.C. § 101 (2006).

[27] Chosun, Int'l, Inc. v. Chrisha Creations, Ltd., 413 F.3d 324, 328 (9th Cir. 2005).

[28] The term "useful article" is defined as "an article having an intrinsic utilitarian function that is not merely to portray the appearance of an article or to convey information." 17 U.S.C. § 101 (2006).

[29] CRS Report RS22685, *Copyright Protection for Fashion Design: A Legal Analysis of the Design Piracy Prohibition Act (H.R. 2196)*, by Brian T. Yeh.

[30] Schechter and Thomas, *supra*, at 587-91.

[31] Wal-Mart Stores, Inc. v. Samara Bros., Inc., 529 U.S. 205, 209 (2000).

[32] Ingrida Karins Berzins, "The Emerging Circuit Split Over Secondary Meaning in Trade Dress Law," 152 *University of Pennsylvania Law Review* (2004), 1661.

[33] 15 U.S.C. § 1051 (2006).

[34] *See* RESTATEMENT (THIRD) OF UNFAIR COMPETITION § 20.

[35] *Id.* at § 30.

[36] *See* Margreth Barrett, "Consolidating the Diffuse Paths to Trade Dress Functionality: Encountering *Traffix* on the Way to *Sears*," 61 *Washington and Lee Law Review* (2004), 79; *see also* Jay Dratler, Jr., "Trademark Protection for Industrial Designs," 1988 *University of Illinois Law Review* (1988), 887.

[37] Schechter and Thomas, *supra*, at 610-11.

[38] *See* Qualitex Co. v. Jacobson Prods. Co., 514 U.S. 159, 164-65 (1995).

[39] 35 U.S.C. § 101 (2006).

[40] 35 U.S.C. § 171 (2006).

[41] Saidman DesignLawGroup, "Design Patent" (available at http://www.designlawgroup.com/ tools _designjatent.cfm).
[42] 35 U.S.C. § 103(a) (2006).
[43] Arminak and Associates, Inc. v. Saint-Gobain Calmer, Inc., 501 F.3d 1314, 1319 (Fed. Cir. 2007).
[44] *Id.*
[45] Keebaugh, *supra,* at 260.
[46] Goodyear Tire & Rubber Co. v. Hercules Tire & Rubber Co., 162 F.3d 1113, 1116-17 (Fed. Cir. 1988); *see also* 35 U.S.C. § 289 (2006).
[47] In re Mann, 861 F.2d 1581, 1582 (Fed. Cir. 1988).
[48] Gorham Co. v. White, 81 U.S. (14 Wall.) 511, 528 (1871).
[49] Litton Sys., Inc. v. Whirlpool Corp., 728 F.2d 1423, 1444 (Fed. Cir. 1984) (citation omitted).
[50] 35 U.S.C. § 173 (2006).
[51] Richard Sybert and Lindsay J. Hulley, "Copyright Protection for 'Useful Articles,'" 54 *Journal of the Copyright Society of the USA* (2007), 419.
[52] Scott D. Locke, "Fifth Avenue and the Patent Lawyer: Strategies for Using Design Patents to Increase the Value of Fashion and Luxury Goods Companies," 5 *John Marshall Review of Intellectual Property Law* (2005), 40.
[53] Daniel H. Brean, "Enough Is Enough: Time to Eliminate Design Patents and Rely on More Appropriate Copyright or Trademark Protection for Product Designs," 16 *Texas Intellectual Property Law Journal* (2008), 325.
[54] Jonathan E. Moskin, "The Shape of Things to Come—Emerging Theories of Design Protection," 92 *The Trademark Reporter* (2003), 681.
[55] Gerard N. Magliocca, "Ornamental Design and Incremental Innovation," 86 *Marquette Law Review* (2003), 852.
[56] Keebaugh, *supra*, at 262.
[57] *See* Steven A. Church, "The Weakening of the Presumption of Validity for Design Patents: Continued Confusion Under the Functionality and Matter of Concern Doctrines," 30 *Indiana Law Review* (1997), 499.
[58] Olson, *supra*, at 178.
[59] Fla.Stat. § 559.94 (1987) (repealed).
[60] 489 U.S. 141 (1989).
[61] The Vessel Hull Design Protection Act was enacted as Title V of the Digital Millennium Copyright Act of 1998 in the 105th Congress. P.L. 304, 112 Stat. 2905 (1998).
[62] 17 U.S.C. §§ 1301(b)(2), 1310(b) (2006).
[63] Keebaugh, *supra*, at 274.
[64] 17 U.S.C. § 1306 (2006).
[65] 17 U.S.C. § 1308 (2006).
[66] 17 U.S.C. § 1302 (2006).
[67] 17 U.S.C. § 1309(e) (2006).
[68] 17 U.S.C. § 1309(b)(1) (2006) (further explaining that purchasing or giving an order to purchase an infringing article in the ordinary course of business does not of itself constitute infringement or collusion).
[69] 17 U.S.C. § 1309(b)(2) (2006).
[70] 17 U.S.C. § 1309(c) (2006).
[71] 17 U.S.C. § 1321 (2006).
[72] 17 U.S.C. § 1322 (2006).
[73] 17 U.S.C. § 1323(a) (2006).
[74] *Id.*
[75] 17 U.S.C. § 1305 (2006).
[76] Olson, *supra*, at 177.
[77] 17 U.S.C. § 1301(b)(4) (2006). The statutory definition expressly excludes "masts, sails, yards, and rigging" from the definition of "hull." *Id.*

[78] *See* Maverick Boat Co. v. American Marine Holdings, Inc., 70 USPQ2d 1493 (2004), *aff'd*, 418 F.3d 1186 (11th Cir. 2005).
[79] Olson, *supra*, at 180.
[80] In particular, the word "hull" is defined in part as "the exterior frame or body of a vessel," while the word "deck" is defined in part as "the horizontal surface of a vessel that covers the hull...."
[81] P.L. 110-434, 110th Cong., 2d. sess., 122 Stat. 4972 (2008).
[82] Magliocca, *supra*, at 845.
[83] *See* Laura C. Marshall, "Catwalk Copycats: Why Congress Should Adopt a Modified Version of the Design Piracy Prohibition Act," 14 *Journal of Intellectual Property Law* (2007), 305.
[84] *See* Testimony of Steve Maiman Before the U.S. House of Representatives Subcommittee on the Courts, the Internet and Intellectual Property (February 14, 2008) (available at http://judiciary
[85] For a more thorough review of these proposals, see CRS Report RS22685, *Copyright Protection for Fashion Design: A Legal Analysis of the Design Piracy Prohibition Act (H.R. 2196)*, by Brian T. Yeh.
[86] H.R. 2196 at § 2(a)(2) (2007).
[87] *Id.*
[88] *Id.* at § 2(f).
[89] *Id.* at § 2(d).
[90] *See* Day, *supra*, at 243.
[91] A Bill to Provide Protection for Fashion Designs: Hearing Before the House Subcommittee on Courts, the Internet, and Intellectual Property, 109th Cong., 2nd sess. (2006) (statement of the U.S. Copyright Office) [hereinafter "Hearings"].
[92] *See supra* notes 31-32 and accompanying text.
[93] *See* Hearings, *supra* (statement of U.S. Copyright Office).
[94] *Id.*
[95] *See* Day, *supra*, at 241-42.
[96] *See* Hearings, *supra* (statement of Susan Scafidi, Associate Professor of Law, Southern Methodist University).
[97] Day, *supra*, at 270.
[98] *See* Hearings, *supra* (statement of Susan Scafidi, Associate Professor of Law, Southern Methodist University).
[99] *See* Hearings, *supra* (statement of David Wolfe, Creative Director, Donegar Creative Services).
[100] *See* Hearings, *supra* (statement of Christopher Sprigman, Associate Professor, University of Virginia School of Law).
[101] *Id.*
[102] World Intellectual Property Organization, Hague Agreement Concerning the International Deposit of Industrial Designs, July 2, 1999, 2279 U.N.T.S. 156 (available at http://www.wipo.int/export/sites/www/hague/en/legal_texts/pdf/ geneva_act_1999.pdf).
[103] *See* Fryer, *supra*, at 662.
[104] *See* CRS Report RL3 1132, *Multinational Patent Acquisition and Enforcement: Public Policy Challenges and Opportunities for Innovative Firms*, by John R. Thomas.
[105] *See* World Intellectual Property Organization, "The Hague Agreement Concerning the International Registration of Industrial Designs: Main Features and Advantages" (available at http://www.wipo.int/freepublications/en/designs/91 1/ wipo_pub_91 1.pdf).
[106] *Id.*
[107] Treaty No. 109-2 1 (December 7, 2007).
[108] Message from the President of the United States Transmitting the Geneva Act to the Hague Convention Concerning the International Registration of Industrial Designs (November 13, 2006).

[109] *See* "Intellectual Property/Treaties: PTO Urges Senate Foreign Relations Committee to Ratify Three IP Treaties," 74 *Patent, Trademark & Copyright Journal* (Bureau of National Affairs July 20, 2007), 346.

[110] *See* Erica Pruetz, "Protecting Car Design Internationally: A Comparison of British and American Design Laws," 24 *Loyola of Los Angeles International and Comparative Law Review* (2002), 475.

[111] For further discussion reflecting different views in this debate, see Kenneth Enborg, "Industrial Design Protection in the Automobile Industry," 19 *University of Baltimore Law Review* (1989-90), 227; James F. Fitzpatrick, "Industrial Design Protection and Competition in Automobile Replacement Parts—Back to Monopoly Profits?," 19 *University of Baltimore Law Review* (1989-90), 233.

[112] European Parliament legislative resolution of 12 December 2007 on the proposal for a directive of the European Parliament and of the Council amending Directive 98/71/EC on the legal protection of designs (available at *http://www.europarl.europa .eu/oeil/file.jsp?id=5201742). See* Joseph Straus, "Design Protection for Spare Parts Gone in Europe? Proposed Changes to the EC Directive: The Commission's Mandate and its Doubtful Execution," 27 *European Intellectual Property Review* (2005), 391.

[113] No. 337-TA-557, 2007 WL 2021234 (ITC May 4, 2007).

[114] *See* "Legislation/Patents: Auto Parts Coalition Asks Lawmakers for Design Patent Infringement Exemption," 74 *Patent, Trademark & Copyright Journal* (Bureau of National Affairs September 21, 2007), 611.

[115] H.R. 3059; S. 1368 (both titled the "Access to Repair Parts Act").

[116] *See* "Legislation/Design Patents: Lofgren Introduces Bill Exempting 'Repair' Parts from Infringement Liability," 75 *Patent, Trademark & Copyright Journal* (Bureau of National Affairs March 21, 2008), 527.

[117] *See* "Patents/Legislation: Labor, Auto Safety Groups Oppose Provisions in Patent Reform Measure," 74 *Patent, Trademark & Copyright Journal* (Bureau of National Affairs August 3, 2007), 405.

[118] *See* "Legislation/Patents: Auto Parts Groups Lobby for Bill to Exempt Repair Parts from Patent Infringement," 75 *Patent, Trademark & Copyright Journal* (Bureau of National Affairs April 25, 2008), 674.

[119] Perry J. Saidman, "The Crisis in the Law of Designs," 89 *Journal of the Patent & Trademark Office Society* (2007), 301.

[120] 501 F.3d 1314 (Fed. Cir. 2007).

[121] Gorham Co. v. White, 81 U.S. (14 Wall.) 511, 528 (1871).

[122] *Id.*

[123] 501 F.3d at 1323.

[124] Saidman, *supra*, at 301.

[125] *Id.* at 314.

[126] 543 F.3d 665 (Fed. Cir. 2008).

[127] *Id.* at 668.

[128] *Id.* As described below, the Court of Appeals ultimately determined that the accused nail buffers did not infringe the asserted design patent.

[129] *See* Kyle J. Fiet, "Restoring the Promise of *Markman*: Interlocutory Patent Appeals Reevaluated Post-*Philips v. AWH Corp.*," 84 *North Carolina Law Review* (2006), 1291 (noting these views).

[130] *See* Locke, *supra*; Perry J. Saidman and Allison Singh, "The Death of *Gorman v. White*: Killing It Softly with *Markman*," 86 *Journal of the Patent and Trademark Office Society* (2004), 792.

[131] 543 F.3d at 679.

[132] *Id.*

[133] *See* Daniel Adam Nadel, "The Elusive Point of Novelty Test Leaves Design Patent Infringement in Limbo: A Critique of *Lawman Armor Corporation v. Winner International, LLC*," 17 *Federal Circuit Bar Journal* (2008), 343.
[134] Litton Sys., Inc. v. Whirlpool Corp., 728 F.2d 1423, 1444 (Fed. Cir. 1984) (citation omitted).
[135] Egyptian Goddess, 498 F.3d at 1357.
[136] *Id.* at 1358.
[137] *See* Aaron Cook, "Points of Novelty, *Lawman Armor*, and the Destruction of Design Patents," 12 *Journal of Technology Law & Policy* (2007), 103.
[138] *See* Amicus Brief of Elite Group, Inc. and Sensio, Inc. in *Egyptian Goddess, Inc. v. Swisa, Inc.*, No. 2006-1562, 2008 WL 545143 (Fed. Cir. 2008).
[139] 543 F.3d at 678.
[140] *See* Robert Scheffel, "Clear as Mud: Federal Circuit Issues *Egyptian Goddess* Decision," *Intellectual Property Alert* (Sept. 22, 2008).
[141] *See* Erin Coe, "Federal Circuit Examines Design Patent Test," *IP Law 360* (June 2, 2008).
[142] Keebaugh, *supra*, at 257.
[143] *See* Rebecca S. Eisenberg, "Patents and the Progress of Science: Exclusive Rights and Experimental Use," 56 *University of Chicago Law Review* 1017 (1989).
[144] *Id.*
[145] *See supra* notes 24-28 (copyright); 36-3 8 (trade dress); 42-44 (design patents) and accompanying text.
[146] *See generally* J.H. Reichman, "Legal Hybrids Between the Patent and Copyright Paradigms," 94 *Columbia Law Review* (1994), 2432.
[147] *See* CRS Report RL33996, *Patent Reform in the 110th Congress: Innovation Issues*, by John R. Thomas and Wendy H. Schacht.
[148] *See generally* Mark D. Janis and Stephen Smith, "Technological Change and the Design of Plant Variety Protection Regimes," 82 *Chicago-Kent Law Review* (2007), 1557.
[149] 98th Congress, P.L. 620, 98 Stat. 3347 (codified as amended at 17 U.S.C. § 901 *et seq.*).
[150] *See* Leon Radomsky, "Sixteen Years After the Passage of the U.S. Semiconductor Chip Protection Act: Is International Protection Working?," 15 *Berkeley Technology Law Journal* (2000), 1049.
[151] *See* Jonathan D. Carpenter, "Intellectual Property: The Overlap Between Utility Patents, Plant Patents, the PVPA, and Trade Secrets and the Limitations on the Overlap," 81 *North Dakota Law Review* (2005), 171.
[152] Daniel H. Brean, "Enough is Enough: Time to Eliminate Design Patents and Rely on More Appropriate Copyright and Trademark Protection for Product Designs," 16 *Texas Intellectual Property Law Journal* (2008), 325.
[153] 35 U.S.C. § 103(a) (2006).
[154] Saidman, *supra*, at 303.
[155] *See* Steven Wilf, "The Making of the Post-War Paradigm in American Intellectual Property Law," 31 *Colombia Journal of Law & the Arts* (2008), 139.
[156] *See* Michael Abramowicz, "Perfecting Patent Prizes," 56 *Vanderbilt Law Review* (2003), 115.
[157] *See* Peter K. Yu, "The International Enclosure Movement," 82 *Indiana Law Journal* (2007), 827.
[158] *See* Feist Publications, Inc. v. Rural Telephone Service Co., 499 U.S. 340, 349 (1991) (explaining that the inability of copyright to protect facts *per se* is not an unforeseen byproduct of the statutory scheme, but rather an essential characteristic of the copyright system).
[159] 35 U.S.C. § 103 (2006).
[160] KSR International Co. v. Teleflex Inc., 127 S.Ct. 1727, 1746 (2007).
[161] *See* Mark D. Janis, "Second Tier Patent Protection," 40 *Harvard International Law Journal* (1999), 151.
[162] *See* Daniel J. Gervais, "The Protection of Databases," 82 *Chicago-Kent Law Review* (2007), 1109.

[163] *See* Thomas G. Field, Jr., "Copyright Protection for Perfumes," 45 *IDEA: The Journal of Law and Technology* (2004), 19.

[164] *See* Pamela Samuelson *et al.*, "A Manifesto Concerning the Legal Protection of Computer Programs," 94 *Columbia Law Review* (1994), 2308; *but see* Jane C. Ginsburg, "Four Reasons and a Paradox: The Manifest Superiority of Copyright Over *Sui Generis* Protection of Computer Software," 94 *Columbia Law Review* (1994), 2559.

[165] *See* Janis, *supra*, at 151.

In: U.S. Innovation Programs
Editors: Allison E. Berger et al.
ISBN: 978-1-61122-809-0

Chapter 6

INNOVATION MEASUREMENT: TRACKING THE STATE OF INNOVATION IN THE AMERICAN ECONOMY

The Advisory Committee on Measuring Innovation in the 21st Century Economy

ACKNOWLEDGMENTS

The members of the Advisory Committee would like to acknowledge the contributions of those who have assisted us in our task.

We are very appreciative of the assistance we have received from U.S. Department of Commerce personnel. We are particularly grateful for the work of Patricia Buckley, the Executive Director of the Advisory Committee and a Senior Economic Advisor to Secretary Gutierrez, and E. R. Anderson, the Federal Designated Officer for the Advisory Committee and Deputy Under Secretary for Economic Affairs.

We have benefited greatly from the guidance of Cynthia A. Glassman, Under Secretary for Economic Affairs, and appreciate the assistance we have received from her staff at the Department as we have prepared this chapter.*

* Particular thanks go to: Jane W. Molloy, David N. Beede, Sabrina L. Montes, Joseph V. Kennedy, Cassandra A. Ingram, Beethika S. Khan, Daniel D. Bachman, David K. Henry, Kemble Stokes, Jacque Mason, and Dar Davis.

We are also grateful for the support of the Department's statistical agencies: the Census Bureau and the Bureau of Economic Analysis.

In addition, we wish to thank all those who submitted comments to us in response to our *Federal Register* request. All of the submissions helped us greatly in thinking through our ideas and formulating our recommendations.

EXECUTIVE SUMMARY

Measuring Innovation and Its Impact on the Economy

The Advisory Committee on Measuring Innovation in the 21st Century Economy was established by the Secretary of Commerce in September 2006 to recommend ways to improve the measurement of innovation in the economy. In its report, the Advisory Committee outlines its recommendations to the Secretary of Commerce for steps to be taken by the government, the business community, and government and private sector researchers to foster and improve the measurement of innovation in the economy.

The first act of the Advisory Committee was to establish a definition of innovation that would identify what should be measured. The definition adopted by the Committee is:

> The design, invention, development and/or implementation of new or altered products, services, processes, systems, organizational structures, or business models for the purpose of creating new value for customers and financial returns for the firm.

The definition recognizes that the innovation to be measured is more than simply something new; it has the added component of adding value for both customers and firms. The definition also recognizes that innovation measurement needs to extend beyond simply measuring inputs. While it is important to track inputs to innovation — such as research and development spending — that is not enough. Outcomes of innovative activity need to be tracked and measured to determine fully the impact of innovation on the economy.

As part of its work, the Committee published a *Federal Register* request for public comments and received many helpful responses. These comments were extremely useful in informing both the guiding principles and the recommendations adopted by the Committee. The *Federal Register* request and copies of all the comments received can be accessed on the Advisory Committee's web site at www.innovationmetrics.gov.

The following set of principles was developed by the Advisory Committee to guide its own work:

- Innovation data collection efforts should build on the way firms assess the effectiveness of their innovative activities.
- When developing better ways to quantify innovation in the marketplace, consideration should be given to measuring the impact of legislation and regulations on innovation.
- Because of the nature of innovation and, in particular, the collaborative nature of the innovative process, there needs to be tolerance of qualitative and subjective measures.
- Innovation measurement should not be static. Measurement is an iterative process that needs to be treated less like a 'project' and more like an ongoing 'dialogue.' Learning and improvement are to be gained from each stage of the process.
- Innovation measures should allow for analysis at the establishment, firm, industry, country, international, and, where possible, regional levels.
- A conservative approach should be taken to any new data collection effort by recognizing tradeoffs between costs and potential benefits and considering resource and regulatory constraints. The implementation of pilot projects to gauge the costs and benefits of new data collection efforts is encouraged. To the extent possible, new innovation measures should be able to be 'back-tested; that is, if applied to historical data, the measures should produce the expected innovation relationship.

The principles guided the work of the Advisory Committee and should also apply to implementation of the Committee's recommendations.

The bulk of the Advisory Committee's work was devoted to developing recommendations to the Secretary of Commerce for actions to improve innovation measurement. The recommendations endorsed by the Committee are summarized briefly below and appear in a more comprehensive list at the end of this summary. All of the recommendations are described in more detail in the body of the report.

What the Government Should Do

To achieve the long-term goal of measuring the impact of innovation on the economy, the Advisory Committee recommends that the government create a coordinated emphasis on innovation measurement. The effort will require structural refinements of existing government data sets, the collection of new and better data, improved linkages among statistical agency data sets, and expanded data sharing/synchronization authority.

In particular, the Advisory Committee recommends that the government:

- Create a stronger framework for identifying and measuring innovation in the national economy.
- Better leverage existing data among the statistical agencies to allow for the consistent estimation of the contributions of innovation in the gross domestic product (GDP) and productivity accounts and to develop greater understanding of innovation.
- Increase access to data in order to facilitate more robust innovation research.
- Convene one or more workshops or forums under the auspices of the Secretary of Commerce to discuss innovation drivers, impediments and enablers.
- Continue participation in the international dialogue related to measuring and analyzing innovation and ensure that U.S. efforts are internationally compatible to the extent possible.
- Consider development of a national innovation index when more work has been done on both data collection and analysis of innovation drivers.
- Support funding necessary to implement the above recommendations.

Most of the recommendations for government action build on existing programs or activities. These include the U.S. National Income and Product Accounts (NIPAs) which constitute the official framework used by the Commerce Department's Bureau of Economic Analysis (BEA) to estimate output, income and expenditure, trade, capital formation, and wealth in the U.S. economy. The NIPAs were devised to help policymakers deal with the severe economic fluctuations produced by the Great Depression and World War II and have been continually refined since then. The NIPAs now provide policymakers with an unrivaled ability to adjust policy quickly and appropriately in response to short-term economic fluctuations. The NIPAs were not originally designed to measure innovation or delve into the causes of long-term productivity growth. Today, however, when policymakers are increasingly shifting attention from short-term stabilization to long-term economic growth, changes to the NIPAs need to be made to accommodate this new focus.

Refining the framework for measuring the performance of the national economy is an essential element in the government's program to measure innovation in the national economy and refine overall economic measurement. These improvements will take time as there is considerable preliminary work to be done not only by BEA but also by the other statistical agencies upon which BEA relies for data. The four major elements of this program are: integrating industry-level estimates of total factor productivity' with the NIPAs; creating a supplemental innovation account; improving service sector data; and improving the measurement of intangibles. Work on each of these four elements has already begun, although additional funding will be required to move forward.

The Advisory Committee also recommends that the statistical agencies pursue an agenda in support of the development of linkages between data sets both to improve data consistency and to provide a richer data base for understanding and explaining innovation. Full implementation of such an agenda would require new legislation. The statistical agencies are currently limited in their authority to share data and that, in turn, affects data consistency.

Linkages between establishment-based data and firm- based data would be particularly useful for tracking and measuring innovation. Most U.S. industry statistics are estimated using establishment-level data as the basic 'building blocks.' Such statistics have been very useful for many purposes, as they combine data for establishments that do approximately the same things. However, many firms own or control more than one establishment, and those establishments may be located in different geographic areas and may be producing different kinds of goods or performing different kinds of services, some of which, such as technology licensing transactions, may be of particular importance to the innovation process. Reassembling establishment-level data into firm-based statistics may lead to better innovation measurements. Furthermore, firm- or establishment- level data from one data source can be augmented by matching them to corresponding data from other sources to obtain a more complete picture of innovation. All of these linked data are understood best when also matched over time to create longitudinal data sets.

In the longitudinal records, the dynamics of business and innovation begin to emerge.

The Advisory Committee recommends that the Secretary support legislation to enable the statistical agencies to undertake expanded data sharing/synchronization activities. In particular, amending the law to expand access to IRS data to additional statistical agencies for the purposes of reconciling the business lists and designing more effective survey business frames would improve our understanding of the U.S. economy.

To encourage more research by non-government researchers, the Advisory Committee recommends that the government encourage innovation research by making public data more transparent through the use of data-tagging or similar methods of making data more user-friendly and by improving access to data through the creation of more public use data files. Such efforts are being undertaken currently by some agencies and the Committee encourages the expansion of such efforts. In addition, the Advisory Committee also recommends the expansion of non- government researcher access to confidential micro- data, including that on business dynamics, while maintaining high standards for confidentiality.

A major issue raised by many of the Committee members was the need to examine innovation drivers, impediments, and enablers. Since the issue was outside the scope of the Committee's mandate, it was not fully explored. However, given the importance of the topic, the Committee calls on the Secretary of Commerce to convene one or more workshops or forums to examine innovation drivers, impediments and enablers.

The Advisory Committee recognizes the importance of the international dialogue on innovation and recommends that it be continued and that efforts be made to ensure that new innovation measures allow for analysis across countries.

Finally, the Advisory Committee recommends support for the additional funding that will be necessary to implement the recommendations.

How the Business Community Can Help

Measuring innovation must be a collaborative process, and there is much that the business community can do to assist and drive improvements in innovation measurement. In particular, the Advisory Committee recommends that the business community:

- Create, expand and assess firm and industry- level measures of innovation and develop best practices for innovation management and accounting.
- Participate in research activities and, as appropriate, make innovation information available to researchers.

One of the guiding principles endorsed by the Advisory Committee was that innovation data collection efforts should build on the way firms assess the effectiveness of their innovative activities. Individual firms, trade associations and other organizations are important partners in developing and testing innovation measures.

Where Research is Needed

While our understanding of innovation has increased over recent years, much more needs to be learned about innovation and its measurement. Government, business, and academic researchers should undertake research — alone and in collaborative efforts — to understand innovation better. In particular, the Committee recommends exploration of the following research areas:

- Identification and assessment of innovation outcome measures.
- Identification of gaps in innovation data and how they might be filled.
- Analysis of relationships between innovation activities and collaboration, innovation performance and firm performance.

Conclusion

The recommendations, if adopted, will go far in setting this nation on a course toward effectively measuring the impact of innovation on the economy. The work is essential to understanding and developing better policies for innovation. The Committee calls on the government, the business community and researchers to work together to improve the understanding and measurement of innovation.

Complete List of Advisory Committee Recommendations

What the Government Should Do

Create a stronger framework for identifying and measuring innovation in the national economy.

- Develop annual, industry-level measures of total factor productivity by restructuring the National Income and Product Accounts (NIPAs) to create a more complete and consistent set of accounts integrated with data generated by other statistical agencies to allow for the consistent estimation of the contributions of innovation to economic growth.
- Create a supplemental innovation account for the NIPAs in order to expand the categories of innovation inputs and allow those inputs to be tracked as they flow between industries.
- Improve service sector data and increase survey coverage to provide the data needed to improve estimates from the integrated GDP/productivity accounts and supplemental innovation account.
- Improve measurement of intangibles, particularly intellectual property, building on work currently under way at the National Science Foundation. Consider the best way in which to collect data on transactions involving key intangible assets such as intellectual property licensing expenditures and revenues. The Commerce Department should also explore whether additional identifying information from patent and trademark applicants might provide useful data.

Better leverage existing data among the government statistical agencies to allow for the consistent estimation of the contributions of innovation in the GDP and productivity accounts and to develop greater understanding of innovation.

- Develop linkages within and between existing data; for example, develop linkages between establishment-based data sets and firm-based data sets to provide both greater consistency in estimations and to provide researchers a broader range of innovation data.

- Develop more robust classification methods; for example, classify firms on the basis of both domestic and international activities.
- Seek expanded interagency data sharing/ synchronization legislative authority in order to improve the quality of innovation measures while balancing data needs with confidentiality protection.

Increase access to data in order to facilitate more robust innovation research.

- Increase the transparency of and access to public data by fostering the use of data tagging and similar processes.
- Create more public use files in order to encourage more non-governmental research.
- Expand non-government researcher access to confidential micro-data, including that on business dynamics, while maintaining high standards for confidentiality.

Convene one or more workshops or forums under the auspices of the Secretary of Commerce to discuss innovation drivers, impediments and enablers.

Continue participation in the international dialogue related to measuring and analyzing innovation and ensure that U.S. efforts are internationally compatible to the extent possible.

Consider development of a national innovation index when more work has been done on both data collection and analysis of innovation drivers.

Support funding necessary to implement the above recommendations.

How The Business Community Can Help

Create, expand and assess firm and industry-level measures of innovation and develop best practices for innovation measurement.

- Institute firm-level measures of innovation to test the correlation of particular measures with known innovation and to measure innovation in the firm and its impact on firm performance. Possible measures might be based on market share or on innovation intensity (e.g., the share of firm revenue attributable to recently introduced products and services).

- Develop and implement best practices in innovation management and accounting.

Participate in research activities and, as appropriate, make innovation information available to researchers.

- Participate in collaborative projects as a means of assembling a broad range of data related to innovation.
- File public reports in XBRL (Extensible Business Reporting Language), a data-tagging format, when it is an option.

Where Research is Needed

Much more needs to be learned about innovation and its measurement. Government, business, and academic researchers should undertake research — alone and in collaborative efforts — to explore the following research areas:

Identification and assessment of innovation outcome measures.

- Assessment of the effectiveness of measures based on market share as innovation measures and the feasibility, cost, and burden of developing such measures.
- Assessment of the feasibility, cost, and burden of developing measures of innovation intensity, including a review of other countries' experience in this area, and consideration of a pilot project.
- Analysis of the qualitative and quantitative impacts of specific innovation drivers, impediments and enablers on innovation outcomes.

Identification of gaps in innovation data and how they might be filled.

- Identification of new data that would be useful in measuring innovation.
- Assessment of the feasibility, cost, and burden of collecting data on intellectual property transactions.
- Identification of ways to overcome gaps and shortcomings in historical empirical measures of intangible investments.

Analysis of relationships between innovation activities and collaboration, innovation performance and firm performance.

- Analysis of the relationship between innovation and occupational employment at firms, using firm-level micro-data.
- Evaluation of whether firms with high innovation intensities perform better than otherwise similar firms with low intensities.
- Assessment of the effect of collaboration on innovation outcomes and identification of the key elements of successful collaborative activities.
- Assessment and analysis of cross-national innovation activities of firms.
- Analysis of publicly filed financial and other data on firms, particularly as the data become more user-friendly, to identify innovative practices and firms.
- Description and explanation of business dynamics and, to the extent possible, analysis of their relationship to innovation using longitudinal business databases.
- Exploration of the use of different sources of available data to determine whether there are correlations between innovative performance of firms in different regulatory environments.

A Message from the Chair

Why Measuring Innovation Matters

In one of his final poems, W.B. Yeats observed, "Measurement began our might." Considering the role innovation has played in the story of America's economic expansion, we might say that innovation is our might. As the world's economies flourish, it will become increasingly easy for our fellow citizens and people across the globe alike to forget that this prosperous economic system we all share—what I have referred to as "entrepreneurial capitalism,"[2]—is largely an American invention. To make this observation, however, is not to write a final chapter. Rather, reflecting on this great achievement—sustained long-term growth and the political stability it has engendered—prompts at least four key questions that are central to America's next economic chapter.

The first is, what is the relationship between innovation and economic growth? Even the most casual of observers must notice processes that yield not only improvements in the goods and services we encounter in our daily lives but also radically new inventions, unimaginable but a few years or months earlier. This is the hallmark of daily economic life. From continuous improvements in computers and service delivery to the evolution of methods to make the human gene respond to medical intervention, mankind has come to expect that next year we will enjoy an economy that produces yet deeper meaning to the simple expectation that things will be "better, faster, cheaper." This is the axiom of innovation, and the vocabulary of growth. If we consider the state of economic knowledge, we can look back on a tremendous accumulated body of insight. Our dedication to understanding economic behavior and the workings of markets has led to what should be appreciated as marvelously practical knowledge. As a result of many years of research, mankind is able to operate within a global milieu where central banking systems, efficient capital markets, and highly specialized labor markets have produced expanding welfare for more and more people.

But to look at the formal statistical record of the United States economy, the vocabulary of innovation is muted if not silent. To appreciate this lacuna, it is important to understand that our system of measurement was conceived largely in response to the Great Depression. Since then, our statistical agencies have been remarkably competent at maintaining and refining public data series. However, the framework of our data infrastructure does not support or reflect the nature of today's dynamic economy—an economy that has undergone profound changes over the past seventy years. Whatever innovation is (and the concept, as we shall see throughout this chapter, presents significant definitional problems), its force in the economy is captured only indirectly.

In an era where capital and labor seem abundant for at least the foreseeable future, we have come to see innovation as the most important avenue to growth.

While obviously important, the nexus between innovation and growth is one of the least understood areas of economic life. One reason is that we have not paid sufficient attention to the question of growth itself and what drives economic growth. Prior to the industrial revolution, growth was an unknown dimension of human experience. It is estimated that for millennia, mankind experienced no growth; centuries came and went with lives changing little. Even more importantly, global wealth remained stagnant—few people became richer and few poorer. If there appeared to be an expansion of wealth in one locale, it usually entailed armed conflict in which static goods or precious metals were wrested away from another family, city, or nation. Since about 1820, America's particular form of entrepreneurial capitalism has presented the most consistent expansion of wealth experienced in history – a twenty-fold increase in living standards. And, it was the wealth and opportunity in the United States that caused nearly all nations to set aside centrally planned socialist economic orders in favor of the dynamic and unpredictable phenomenon of one form or another of capitalism.[3]

As suggested, however, economic growth remains something of a mystery The chart that follows presents an illustration of the relationship between economic growth and economic knowledge. The triangle, known for Professor Arnold Harberger, who first observed the limits of economic research, circumscribes most of what has occupied economists over the decades. Studies in monopoly theory, labor, public finance, industrial organization, etc., have a shared goal, namely, advancing the optimized efficiency of the economy so as to capture a marginal gain in growth. But, the story of the growth that counts, the tremendous shift over time of the nation's total production, is largely unaccounted for in such optimization theories. Decades ago, Joseph Schumpeter, the first economist to devote himself to entrepreneurship and innovation, wrote: "This historic and irreversible change in the way of doing things we call 'innovation' and we define: innovations are changes in production functions which cannot be decomposed into infinitesimal steps. Add as many mail-coaches as you please; you will never get a railroad by so doing." Many economists have documented that the most important story of economics is that of how innovation came to be and how it is the hinge that opens the door to economic growth.[4]

The second question that must be addressed in thinking about America's economic future is, what is innovation? In previous ages, invention was thought to be the key to explaining growth in the American economy. We thought of ourselves as a nation marked by our propensity to invent new things and new processes. Over time, our history suggested to us that this was in our character. Frederick Jackson Turner, one of our most distinguished historians, developed the frontier hypothesis that suggested that while Americans thought that exploring and moving west was our defining frontier, this was really only a metaphor for the creative pioneering that Americans feel is one of our distinctive shared attributes.

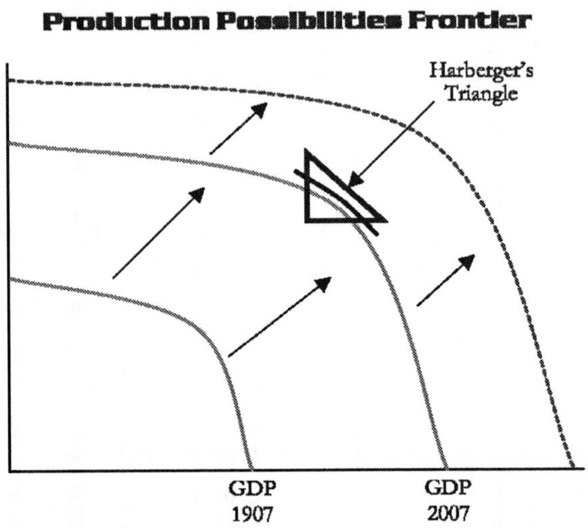

Today we increasingly think of innovation, rather than invention, as a word of our times. We believe our propensity to innovate is central to our character and our economy. But just what is innovation, and how is it different from other phenomena that have been with us since our founding? Obviously, as we think of innovation in the context of this Committee, it is not an ephemeral national characteristic or personality attribute—it is a defining part of our economy. It is the edge where the "new" comes into being and is transformed into a concrete reality that produces benefits. So, one of the first acts of the Committee was to develop a definition of innovation that tied "new" to "impact." The Committee decided that innovation is "the design, invention, development and/or implementation of new or altered products, services, processes, systems, organizational structures, or business models for the purpose of creating new value for customers in a way that improves the financial returns for the firm."

Considering how to define innovation prompts the third question, which is the focus of this chapter—that is, how do we measure innovation? For those who worry about questions related to expanding human welfare through economic growth, it may be the most important practical inquiry of our times. This chapter suggests the complexity of this task. It describes the issues surrounding the definition of innovation. And, as might be expected, it points to what amount to proxy measures. As we deliberated, it became clear that sufficient research does not exist to guide us toward a single measure. Perhaps this will always be the case. The very nature of innovation suggests that it will never yield to a tidy and static metric. An enormously complex economy produces millions of ways in which "better, faster, cheaper" comes into play every single day.

In this regard, the report serves as a pointer, describing the first steps of what will necessarily be a very long, maybe never-ending journey. The Committee recommendations start with a suggestion that better measures of the growth in "total factor productivity" – the change in productivity left over after taking account of the growth of capital and labor – are required to begin to zero in on the contribution of innovation. In reaching this conclusion we have benefited from the particular expertise of Professor Dale Jorgenson, one of our fellow members. But, total factor productivity is, as we've noted, a surrogate approach. Such a measure only accounts for a portion, albeit a large portion, of the change in the economy's performance that might be said to be innovation. There are other measures that add more texture to the task but make any approach to an inclusive measure that much more difficult.

In fact, after extensive consideration, the Committee has concluded that as much as an aggregate measure of the entire economy's innovation might be desirable, recommending an "innovation index," a sort of all-in-one measure of innovation, would be unwise given the current state of research on innovation and economic growth. A single index would be hard to construct and harder still to defend. Innovation being innovation, it would hardly be established before it would have to be changed. Moreover, the economy does not innovate evenly:

> there will always be some firms and sectors that innovate at a greater pace than others. Creating a single index that treats innovation as a single phenomenon might lend itself to policy distortions. It would be used immediately in discussions of what policy steps might be appropriate to stimulate innovation, and an error in the construction of such a unitary index could play through to disastrous consequences.

In the absence of a single indicator, the Committee proposes that the Department of Commerce take steps toward improving, integrating, and expanding on its current data collection efforts in the next few years, while research is undertaken to better inform the task of measuring innovation. Our statistical system captures two of the major inputs that are linked to the innovation process: research and development spending and the number of engineers, scientists, and technicians employed. In addition, data are collected on some other categories of investments, such as expenditures on information technology equipment, which is certainly a factor in expanding innovation. We also measure the number of innovations that are protected by newly-issued patents each year. But, in many firms and industries, significant amounts of investments in innovation are made outside of these categories and go consistently unmeasured or unconnected by the current statistical system. Indeed, even for the variables we measure, such as research and development, the Committee recognizes the need to update the way we quantify these measurements, the frequency of our measurement, and our coverage of younger firms and emerging industries. Further, we must develop more finely calibrated measures of how firms invest in the inputs that become innovation, especially human capital.

In the absence of common public measures of the scope of innovation, many surrogates have sprung up in the private sector. Numerous interest groups, industry associations, and think tanks produce indices of innovation that should be considered as part of the expanding mosaic of data sources. In fact, one of the innovations resulting from this chapter might be that the government absorbs several of the most statistically valid private sector data series into its own overall measures of innovation.

The final question is the most important one of all. What can we do to drive the American economy to a state of continuous and sustained growth by making our country more innovative? Making innovation an explicit concern of public policy requires that we can measure how innovative the economy is, as well as determine whether the "state" of innovation is improving or decaying. It also requires that we understand the pathways of innovation lest we develop the wrong policy interventions, either to quicken innovation or to shift its course. While it is difficult to say what policy should be, some things are understood and should inform policy considerations.

For example, innovation is meaningless if not considered in the larger context of growth. Indeed, growth should be the touchstone of policy, and in the case of innovation, it is important to understand its role. Innovation propels growth by pushing existing companies to real cost reductions and new firms toward growth. The United States remains unique in its ability to grow companies. Every year roughly thirty-one firms achieve "top line" revenue in excess of $1 billion.' Each one of the firms that experiences this kind of growth takes as its basis some innovation in the nature of its products or in the nature of its approach to marketing.

Entrepreneurial activity, both in start-ups and in existing large-scale companies, is critical to the firm's success. Every entrepreneur is a party to innovation; in fact it could be said that entrepreneurs are to our economy as the marines are to our armed forces. Calculated risk taking, with a goal that is often only defined as "let's win," is the mindset of entrepreneurs, both those working on their own and those who shoulder the path-breaking and often risky task of changing the culture and identity of mature firms. To make the economy more congenial to innovation is to construct an economic culture that appreciates what entrepreneurs do.

As we consider these observations, it is clear that our economic ecosystem can be helpful to innovation and its ability to flourish. Policy should first focus on putting in place improvements to our measurement of innovation and then focus on what we know already as key factors for innovation. Without doubt, we must have people who are skilled and trained to see the opportunities where the "better, faster, cheaper" can be realized. This means in real terms that our schools, community colleges, and university-level institutions must prepare more of our young citizens to be better able to advance technical insights such that innovative products and services continue to flow into our economy.

If we are to grow, we need more and more people ready to take entrepreneurial risk. And, we need them flowing from institutions attuned to producing particularly creative people for the new economy. Recent economic and psychological research has confirmed what scientists and entrepreneurs have known for decades: innovative breakthroughs frequently come at the estuary region where different fields, not necessarily related, intersect. This means we need much more cross-disciplinary training where the edge between fields can be developed for the innovations that lie within.

In conclusion, the work of the Committee has brought together experts from across government, industry, and academia to consider changes that can be made today to improve our understanding of the American economy and its innovative capacity in the future. Measuring innovation is central to understanding the economy as it evolves and responds to growing world competition. Indeed, it is better to travel an illuminated path toward future economic progress than to stumble in an unlit direction. Improvements to our measurement of innovation will help to ensure continued economic strength.

1. The Current State of Innovation Measurement and the Establishment of the Advisory Committee

Innovation propels economic growth. Yet the measurement of innovation, in this country and elsewhere, remains rudimentary To understand better the dynamics of economic growth and, hopefully, avoid harmful policies and enact facilitative policies, we must design improved measures of innovation. With this objective in mind, the Secretary of Commerce, Carlos M. Gutierrez, established the Advisory Committee on Measuring Innovation in the 21st Century Economy.

In the popular mind, innovation is often equated with invention. In some cases, this is correct – Thomas Edison stands out as a prominent example – but invention is simply one type of innovation. Innovation can be an astounding breakthrough (as was often the case with Edison), a mundane shift in process, or a subtle change in culture. Yet each type helps generate a higher yield on resources, boosting economic growth. Examples provided by members of the Advisory Committee at its first meeting illustrate this variety.

- David Bernd, CEO of Sentara Healthcare, an integrated health care provider in Virginia, spoke of getting a cold call from a small start-up company. The caller told him about an innovative method of remotely monitoring intensive care beds from a centralized location. Sentara ultimately became the first hospital to use the system, and now remotely monitors 101 intensive care beds in five institutions. The hope was that the new system would improve the quality of care. At the end of two years, Sentara found that mortality had been reduced by 15 percent, ICU length of stay was reduced by 16 percent, variable costs went down 25 percent and retention of registered nurses in the covered units went up 20 percent.
- James Blanchard, former CEO of Synovus, spoke of the creation of his company's subsidiary Total System Services, now one of the world's largest processors of credit cards. It started when his small Georgia bank, one of the first to issue a credit card, bought some computers, wrote a program from scratch and automated the entire operation. Ten years later, BankAmericard automated and Synovus became the first bank to use the process. In 1974, almost as a joke, the company offered to process cards for another bank. The other bank

accepted the offer and that started a process that continues today. In 1983 Synovus spun off the operation, capitalized it, and did an initial public offering (IPO). Today it is listed on the New York Stock Exchange and processes over 400 million credit card accounts all over the world.

- John Menzer, Vice Chair of Wal-Mart, spoke of his company's process-based gains in the areas of environmental sustainability, prescription drugs and supply chain efficiencies. He considers Wal-Mart's process innovation to be a unique marriage of corporate culture, imagination and technology. He spoke of setting big stretch goals for the company and associates—a 25 percent more efficient trucking fleet in three years and double in ten years, a 20 percent reduction in energy use in new stores in four years, and a 25 percent reduction in solid waste in three years. The company has no specific plans in place to meet the goals but uses the goals to inspire associates to help the company get there, create a business environment that thrives on new ideas and change, and challenge their supplier base to set individual goals. And he noted that all Wal-Mart's innovation is geared toward a better consumer experience and better returns for the company.
- Michael Eskew, Chairman and CEO of UPS, spoke about how UPS transformed itself from a local delivery company, to a ground company, to an air company, to an international company, to a supply chain company and to what is now largely a technology company. UPS currently uses four innovation processes. For innovation related to its core business it relies both on an internal marketing committee and on external sources, primarily customers. For 'non-core' innovation, UPS uses both an internal strategy group and external groups including universities and pre-IPO venture companies funded by a UPS strategic enterprise fund.

These examples underscore the challenges in attempting to measure innovation in the economy.

The State of Innovation Measurement

Innovation measurement is in its infancy — both here and around the world. Some data related to innovation are collected by the U.S. government statistical agencies; however, the data are incomplete and miss substantial sources of innovation in the economy. Some innovation data are tracked by private sector organizations and firms, but these efforts are also limited. Recently, however, government agencies, businesses and trade associations have been devoting more resources toward the development of such data.

Examples of some of the U.S. statistical agency programs that either currently collect innovation- related data or could be vehicles for enhanced innovation data collection and analysis through linkages, new data or new analysis include the following:

- The Commerce Department's Bureau of Economic Analysis produces the U. S. National Income and Product Accounts (NIPAs). The major agencies providing data for the NIPAs include: (1) the Census Bureau (data from business and population censuses and surveys); (2) the Bureau of Labor Statistics (employment statistics, wage and salary data, productivity statistics, and most underlying price information); (3) the Internal Revenue Service (tax-based data on individuals and businesses); and (4) the Federal Reserve Board (income statements and balance sheets for major financial and non-financial sectors).
- The Bureau of Economic Analysis, with support from the National Science Foundation (NSF), has developed a Research and Development Satellite Account that estimates the effect of investment in research and development on U.S. economic growth. These experimental estimates of the effect of intangible assets on the U.S. GDP show the size of the impact of research and development (R&D) on U.S. economic growth. BEA also collects data on international technology licensing expenditures.
- The Census Bureau has data on some measures of innovative activities, such as the diffusion of innovation, human and organizational capital, entrepreneurship and other worker and firm characteristics. It is working on improvements in some areas of importance to innovation, particularly related to the service sector. Detailed information has long been collected on manufacturing activities; only recently has emphasis been put on non-manufacturing sectors. The Census Bureau, on behalf of the National Science Foundation (NSF), conducts an annual firm-level survey of industrial R&D that requests data on firms' R&D investments. Census has also created an Integrated Longitudinal Business Database (ILBD) that

includes businesses with and without employees, permitting tracking of the growth of start-ups and other aspects of business dynamics.
- The Securities and Exchange Commission (SEC) collects extensive data from public companies. The SEC has embarked on a data-tagging project that will permit analysis of such financial data across public firms.

These examples represent just some of the sources of innovation-related data available. While they currently provide data on only some pieces of the innovation puzzle, they provide a trove of possibilities for the development of innovation data in the future.

Dedicated innovation surveys such as the European Community's Community Innovation Survey are used by statistical organizations in European Union and some other countries including Australia and Canada. Relatively new, and only tested among manufacturing firms in the U.S., the surveys collect information on different varieties of innovations, including 'new to the firm, new to the industry and new to the world.' They also collect extensive information on innovation expenditures (e.g., capital investment, training and marketing costs) and costs of protecting innovation (e.g., patent and copyright costs). However, such surveys are very costly and have encountered both definitional and response rate problems.

The Establishment of the Advisory Committee

Secretary Gutierrez formed the Advisory Committee to enlist the help of leading business representatives and academics in formulating a plan for improving innovation measurement in this country. One of the first steps taken by the Committee was the adoption of a definition.

For the purposes of its work, the Advisory Committee defined innovation as:

> The design, invention, development and/or implementation of new or altered products, services, processes, systems, organizational structures, or business models for the purpose of creating new value for customers and financial returns for the firm.

Early last year, the Advisory Committee published a *Federal Register* request for public comments on innovation measurement. The purpose of the notice was to get broad input into the work of the Committee.

In the notice, the Committee asked for comments on improved or new innovation measures. The Committee sought comments in four categories:

(1) improvement of the underlying architecture of the U.S. system of national accounts to facilitate development of an improved and more detailed measure of innovation and productivity;
(2) identification of appropriate economy-wide and sector-specific statistical series or other indicators;
(3) identification of firm-specific data items that could enable comparisons and aggregation; and
(4) identification of specific 'holes' in the current data collection system that limit our ability to measure innovation.

The Committee received a broad array of thoughtful and substantive responses that covered the full range of categories for which comments were sought. All of the comments were considered during the Committee's deliberations.

In determining the path on which to direct its recommendations, the Advisory Committee determined that its recommendations would build on existing and ongoing work to the extent possible. Where appropriate, new approaches or new data would be recommended but, in the interests of cost and delivery time, the primary focus would be on improving the scope and robustness of existing work.

The Committee recognized that, at least at this time, there is only limited knowledge about innovation drivers, impediments and enablers. And there was awareness of the fact that many businesses do not currently collect all the data that might be desired in an ideal world.

Because it is also clear that innovation measurement is still in its infancy, the Committee chose to recommend differing approaches to innovation measurement. There is no single innovation measure that can be recommended at this time, but the data and analysis to be generated by the improved innovation measures recommended by the Committee should lead to greater understanding of the process of innovation and prove valuable to policymakers in the future.

The next chapters detail the principles and recommendations endorsed by the Advisory Committee.

2. GUIDING PRINCIPLES FOR INNOVATION MEASUREMENT

The Advisory Committee's recommendations are intended to improve the measurement of innovation and its impact on the U.S. economy. In particular, the goal is to develop better estimates of the resources devoted to and value generated by innovation by focusing not only on measuring innovation activities and inputs but also on innovation results and outputs.

Measuring innovation has not been a major goal of economic data collection by U.S. statistical agencies. Innovation measurement to date has been largely piecemeal, incomplete, and accidental (i.e., relying on data sets not designed for the purpose of innovation measurement).

Improving our understanding of how much is spent on innovation and how much we benefit from innovation is crucial to answering several key questions: Are we as a society spending too much or too little on innovation? Are the places in which focus is being given to innovation having results? And what, if anything, can we do to improve our innovative activities in the future?

The Committee's recommendations call on the government, the business community and researchers (in both government and the private sector) to work together to further both the understanding and the measurement of innovation in the economy.

The Advisory Committee recommends that the work be guided by the following principles.

- *Innovation data collection efforts should build on the way firms assess the effectiveness of their innovative activities.*

Data collection should be informed by what firms are actually doing rather than being based solely on theory. Also, to the extent possible, the burden on firms should be minimized.

- *When developing better ways to quantify innovation in the marketplace, consideration should be given to measuring the impact of legislation and regulations on innovation.*

Some regulatory policies (e.g., certain tax and education policies) may explicitly support innovation. Other policies may have the unintended consequence of inhibiting innovation (e.g., overly restrictive caps on immigration and some SarbanesOxley rules). Improved data on innovation are essential for assessing the impact that regulatory policies have on innovation.

Better firm-specific and economy-wide data will help reveal the impact — both positive and negative — of regulatory policies on innovation and the environment that produces innovation.

- *Because of the nature of innovation and; in particular, the collaborative nature of the innovative process, there needs to be tolerance of qualitative and subjective measures.*

Not all measures of innovation may be quantifiable; and progress in developing better quantification of some dimensions of innovation may depend on improved qualitative measures. For example, measuring the resources invested in and the outcomes of collaboration may be very important but also very difficult, especially if such partnerships are informal or if the benefits are subject to spillovers (i.e., are difficult to capture fully in a contractual arrangement).

- *Innovation measurement should not be static. Measurement is an iterative process that needs to be treated less like a project' and more like an ongoing 'dialogue.' Learning and improvement are to be gained from each stage of the process.*

As new innovation data are collected, they should be refined and continually re-evaluated for their cost- effectiveness and ability to push out the frontiers of knowledge about innovation and its impact on the economy. The government needs the aid of the researcher community in order to do so in a timely manner. And while timeliness is important in understanding innovation, different measures of innovation may capture relatively quick-return innovation while other measures may capture innovations which require longer periods of time to cause measurable economic change.

- *Innovation measures should allow for analysis at the establishment, firm, industry, country, international and where possible, regional levels.*

Improved data on innovation should permit industry- and sector-specific analysis, recognizing that innovation manifests itself differently in different parts of the economy. In particular, international comparisons would help explain why different countries are experiencing different economic growth rates. When compared with different national policy mixes, it may be possible to achieve a better understanding of the impact public policies have on growth and innovation.

- *A conservative approach should be taken to any new data collection efforts by recognizing tradeoffi between costs and potential benefits and considering resource and regulatory constraints. The implementation of pilot projects to gauge the costs and benefits of new data collection efforts is encouraged To the extent possible, new innovation measures should be able to be `back-tested; that is, if applied to historical data, the measures should produce the expected innovation relationship.*

The costs of new data collection include both direct program costs and the cost burden imposed on potential survey respondents. Many conceptual ideas for new data collection need to be tested on diverse industry samples of large and small or young and old firms before new surveys (or substantial changes to existing surveys) are implemented.

3. What Government Should Do

To measure innovation and its economic impact better, the government must make structural refinements to existing government data sets, improve linkages among them, collect new and better data, and expand data sharing/synchronization.

The Advisory Committee's recommendations for government action and the elements within each recommendation are discussed in this chapter.

1. The Government Should Create a Stronger Framework for Identifying and Measuring Innovation in the National Economy

Refining the framework for measuring the performance of the national economy is an essential element in the government's program to measure innovation. These improvements will take time, as there is considerable preliminary work to be done. The four major elements of this program include: refining the National Income and Product Accounts (NIPAs) to permit estimation of industry-level measures of total factor productivity; creating an innovation supplementary account; improving service sector data; and improving the measurement of intangibles (particularly intellectual property and improved measures of technology licensing activity). Each of these elements is discussed in more detail below.

- *Develop annual, industry-level measures of total factor productivity by restructuring the NIPAs to create a more complete and consistent set of accounts integrated with data generated by other statistical agencies to allow for the consistent estimation of the contribution of innovation to economic growth.*

The NIPAs constitute the official framework used by the Commerce Department's Bureau of Economic Analysis (BEA) to estimate output, income and expenditure, trade, capital formation, and wealth in the U.S. economy. The NIPAs were devised to help policymakers deal with the severe economic fluctuations produced by the Great Depression and World War II. The NIPAs were not originally designed to measure innovation or delve into the causes of long-term productivity growth. Today, however, when policymakers are increasingly shifting attention from short-term stabilization to long-term economic growth, changes need to be made to measure innovation in this context.

Economic growth depends — in large measure — on productivity growth. Innovation is a major determinant of productivity growth. Without innovation, output generally can only grow by increasing inputs — through a combination of expanding the labor force and increasing the utilization of capital (such as machines, buildings, and skills, using existing technologies). The implementation by entrepreneurs, managers, and employees of innovations — in the form of new products or services, processes, organizational structures, or business models — enables output growth to exceed the growth of inputs. Measuring that excess growth — known as Total Factor Productivity (TFP) growth — for the entire private sector economy as well as by industry is the primary goal of this proposal. A longer-term goal, discussed later in this chapter, is to parse TFP growth estimates into amounts attributable to expenditures by firms on scientific R&D and information technologies (using available data), and other measurable investments by firms in innovation (using data the Committee proposes to be collected), and amounts due to other factors.

The Committee recommends an ambitious program to eliminate gaps and inconsistencies within the NIPAs and between data produced by other statistical agencies and those produced by BEA. The proposed new 'architecture' for the NIPAs would consist of a set of income statements, balance sheets, flow of funds statements, and productivity estimates for the entire economy and by sector that are more accurate and internally consistent. The new architecture will make the NIPAs much more relevant to today's technology-driven and globalizing economy and will facilitate the publication of much more detailed and reliable estimates of innovation's contribution to productivity growth.

The most salient (for purposes of this Committee) goal of the proposal is to integrate BLS's current TFP (also known as `multifactor productivity') growth estimation program with the NIPAs and extend their coverage to all industries. Doing so would require completion of BEA's plan to improve its industry- level statistics, including fully integrating its input- output tables with the NIPAs and incorporating new data collected by the Census Bureau on the output and intermediate inputs of services industries, as described later in this chapter. Greater coordination between BEA and BLS will be required in other areas as well, including increased data sharing between the two agencies; acceleration of the development of the capital services accounts; and expansion of efforts to incorporate quality adjustments and identify new products and services for prompt inclusion in BLS's Consumer, Wholesale, and International Price Index programs. In addition, BEA will have to work with the Federal Reserve Board (FRB) to develop an integrated wealth account.

The new NIPA architecture will be flexible enough to accommodate the inclusion of information from new supplemental accounts and to permit the easy incorporation of these accounts into the proposed NIPA system. Data developed for capital stocks and service flows for intangible assets (such as technology that passes among firms through licensing agreements and other forms of intellectual property) can be used to parse TFP growth estimates.

The TFP growth estimates proposed by the Committee would be consistent with estimates currently published by the European Union, Canada, Japan, and South Korea and with estimates planned by other countries, including Brazil, China, India, and Russia. The availability of consistently estimated measures of TFP will allow for rigorous comparisons of the sources of growth across countries and over time. Such information would be valuable to policymakers trying to gauge the effectiveness of national innovation policies.

- *Create a supplemental innovation account for the National Income and Product Accounts (NIPAs) in order to expand the categories of innovation inputs and allow those inputs to be tracked as they flow between industries.*

Supplemental accounts (sometimes referred to as satellite accounts) provide additional detail about a part of the economy using the same structure as the National Income and Product Accounts without being integrated with the NIPAs. Establishing a new supplemental account that is conceptually consistent with the NIPAs but based on data that are not as time-tested as those used in the core NIPAs helps facilitate the development of the NIPAs. For example, in 2006, BEA inaugurated a supplemental account for investments in research and development activities conducted by scientists and engineers; ultimately, BEA proposes to incorporate the R&D supplemental account into the NIPAs.

Developing a broad supplemental innovation account for intangible assets — including estimates of intangible asset capital stocks and service flows — would introduce a broader range of capital inputs into the NIPA product and capital accounts. Such inputs have hitherto been classified as expenses rather than as accumulated capital. Data on intangible asset capital stocks and service flows could be used to further refine estimates of the impacts of innovation on economic growth.

The Committee calls for the development of a new NIPA supplemental innovation account that would include intangible assets, such as investments by firms in research and development (conducted by scientists and engineers as well as workers in other occupations); human capital; patents and trade secrets; copyrights, trademarks and brands; and other forms of intellectual property. The development of a supplemental innovation account depends on the collection of additional data on intangible innovation assets.

- *Improve service sector data and increase survey coverage to provide the data needed to improve estimates from the integrated GDP/productivity accounts and supplemental innovation account.*

It is widely believed that much innovation in recent decades has taken place in the service sector, and this belief is bolstered by evidence (albeit based on incomplete and highly aggregated data) that service sector productivity has increased after a long period of stagnation. However, collection of annual data on the service sector has lagged that of other sectors (in particular, manufacturing), and improved service sector data collection is a necessary step toward better measurement of innovation and its impacts in the service sector.

The Census Bureau has long covered retail and wholesale trade in its annual surveys. Of the rest of the service sector, which currently accounts for 55 percent of GDP, the Census Bureau's annual survey covers only 30 percent of GDP. Put another way, the Census Bureau does not collect annual data on one- quarter of the nation's economy.

The Committee recognizes that the funds needed to implement the Census Bureau's proposal to extend annual survey coverage to all remaining service sectors by FY2010 and to improve the data on service industry inputs have not been appropriated. The current lack of funding will impede some activities necessary for improved innovation measurement. For example, BEA's efforts to develop more detailed data on the input/output matrices to track innovation across industries hinges on the Census Bureau collecting more comprehensive service sector data.

The Committee also calls for the development of a more nuanced classification system with finer granularity of data on different types of economic activity in the service sector. In particular, the Committee calls for improved accounting for transfers of intangible assets (such as intellectual property licensing and assignments) and the development of more detailed and quality-adjusted price indices for services.

Since a number of service sector industries (e.g.,software, consulting, and intellectual property licensing and transfers) are considered particularly innovative, better data on the service sector will improve our understanding of innovation processes. Improved services price indices will improve estimates of inflation-adjusted inputs purchased from the service sector inputs, including those purchased from highly innovative industries, and will enhance the quality of estimates of productivity growth attributable to innovation.

- *Improve measurement of intangibles, particularly intellectual property, building on work curently under way at the National Science Foundation. Consider the best way in which to collect data on transactions involving key intangible assets such as intellectual property licensing expenditures and revenues. The Commerce Department should also explore whether additional identifying information from patent and trademark applicants might provide useful data.*

Statistical agencies largely focus on collecting data on firms' investments in physical capital. Little is known about firms' investments in intangible assets beyond data on expenditures on scientific and engineering research and development collected by the Census Bureau for the National Science Foundation (NSF) and data on own-account software development collected by the Census Bureau. Data on investments by firms in 'research and development' activities *other* than scientific and engineering R&D are scarce, as are data on investments in organizational and human capital and marketing and brand equity.

Data on patent and trademark applications and assignments are publicly available, but it is difficult for researchers to match these data to data on applicants and assignees. Better matching would help researchers develop ways to put a value on intellectual property.

Data are similarly scarce on the income earned on intellectual property. The Bureau of Economic Analysis collects data on cross-border royalty payments for intangible assets, but no data are currently collected on purely domestic royalty revenues and payments.

The lack of information on intangible assets' depreciation rates and how to adjust the value of investments in intangibles for price inflation makes it difficult to estimate the wealth that is created by firms' investments in innovation. Sound estimates of intangible capital stocks are needed to refine estimates of innovation's contribution to productivity growth. However, intangible asset valuation is difficult because of the relative scarcity of directly observable prices, and the data collection and methodological challenges are daunting.

The Committee recommends improvements in intangibles measurement by expanding data collection on intangible investments beyond scientific and engineering R&D and own-account software development to include *all* resources devoted to innovation. Such improvements might be achieved by expanding NSF's R&D survey to encompass a broader range of innovation investment activities. Past experience with collecting such data at BLS and in other countries (notably Canada) could be used to inform the development of a survey in the United States.

The U.S. Patent and Trademark Office should collect additional identifying information from patent and trademark applicants to facilitate matching application and assignment data to census and other data sets. Such information could include Employer Identification Numbers (with appropriate confidentiality safeguards), CUSIP numbers, DUNS numbers, or other widely used firm identifiers. Collecting such additional information should impose little additional burden on patent and trademark applicants and would aid the development of patent quality indices using stock market valuations and other data.

The Committee further recommends that BEA and the Census Bureau collaborate to collect domestic (as well as cross-border) data on revenues from and expenses associated with licensing and transferring technology (patents and trade secrets), copyrights, franchise fees, and trademarks.

Finally, the Committee calls for the development of improved estimates of depreciation rates of intangible assets and specific price indices for intangible assets.

2. The Government Should Better Leverage Existing Data Among the Government Statistical Agencies to Allow for the Consistent Estimation of the Contributions of Innovation in the GDP and Productivity Accounts and to Develop Greater Understanding of Innovation

Much could be learned about innovation from data that have already been collected if the statistical agencies were to develop new linkages. Much of this work would require new legislative authority.

- *Develop linkages within and between existing data; for example, develop linkages between establishment-based data sets and firm-based data sets to provide both greater consistency in estimations and to provide researchers a broader range of innovation data.*

Most U.S. industry statistics are estimated using establishment-level data as the basic 'building blocks.' An establishment is a commercial or non-profit entity at a single physical location that produces goods or performs services, for example, a store or a manufacturing plant. Statistical agencies gather data from establishments located within the United States operating within the same industry and publish descriptive statistics for each industry based on such data. Such statistics have been very useful for many purposes, as they combine data for establishments that do approximately the same things.

However, many firms own or control more than one establishment, and those establishments may be located in different geographic areas and may be producing different kinds of goods or performing different kinds of services, some of which may be important to innovation. Statistics based on firm rather than establishment data can be useful in other contexts, especially in innovation measurement. By disassembling the establishment building blocks used to construct industry statistics and reassembling them into firm-based statistics, better innovation measurements may be obtained. Furthermore, firm or establishment data from one data source can be augmented by matching to corresponding data from other sources to obtain a more complete picture of innovation.

- *Develop more robust classification methods; for example, classify firms on the basis of both domestic and international activities.*

Even in those cases where data are collected and classified at the firm level, classification is based only on the activities of domestic establishments. Consider a firm that designs and markets shoes under its own brand name in the United States but manufactures them abroad. Based exclusively on the data collected from establishments located in the United States, the firm would be classified as a wholesaler, but such a firm would likely be very different in terms of innovation activities from a firm that acted solely as a merchant wholesaler (i.e., buying shoes from manufacturers and reselling them to retailers without any design or manufacturing activities). Classification on the basis of both domestic and international activities would improve innovation measurement.

- *Seek expanded interagency data sharing/ synchronization legislative authority in order to improve the quality of innovation measures while balancing data needs with confidentiality protection.*

While the Committee has given much consideration to new or expanded surveys of firms, much remains to be learned from data that have already been collected. However, because of a host of laws and other restrictions affecting firm micro-data (that is, establishment- or firm-specific data records), agencies are often limited in their abilities to share data with other agencies.

Within the federal government, limitations placed upon the use of IRS data affect the accuracy of business registers used by federal agencies to carry out business surveys. Partially because of these inaccuracies, differences arise in how different agencies measure economic activity. Additionally, existing topical surveys have been hampered by lack of access to a good business frame from which to draw a sample.

The Committee recommends that Congress enact legislation to enable and encourage statistical agencies to expand data sharing activities subject to strict confidentiality requirements. Specifically, the Committee recommends that access to IRS data be expanded to additional statistical agencies for the purposes of reconciling the business lists and designing effective special surveys. The grant of such authority would greatly enhance data quality and consistency and improve innovation measurement.

3. The Government Should Increase Access to Data in Order to Facilitate More Robust Innovation Research

- *Increase the transparency of and access to public data by fostering the use of data tagging and similar processes.*

The government collects a vast store of non- confidential data from firms. Analysis of such data by researchers might provide considerable new information on innovation and innovation-related activities. For example, analysis of financial data and reports filed with the Securities and Exchange Commission (SEC) by public companies might provide insights into what differentiates innovative firms. But, until recently, such data have been locked into a format consisting of fixed blocks of text, making it difficult for researchers to process and analyze the data.

Government agencies have begun a process of making public data more user-friendly; for example, the SEC, the Federal Deposit Insurance Corporation and other banking regulators, and several European governments are using or testing XBRL (Extensible Business Reporting Language), a process that 'tags' data (i.e., attaches a standardized, searchable label to each data item filed by firms). Tagging enables `intelligent' computerized recognition of each piece of information in a document, thereby facilitating accurate, automated, and flexible selection, analysis, storage, exchange, and presentation of data at high speed and reduced cost. The SEC data-tagging project is still in its pilot stage but some firms are reporting tagged data.

Expanded use of data tagging of public information could increase transparency of public information and provide a wealth of information for researchers and other members of the public.

- *Create more public use files in order to encourage more non-governmental research.*

Statistical agencies should consider creating more public use files that would be accessible to non- government researchers. While confidentiality requirements may impose some limits to the depth of public use files, greater public access through web- based public use files could increase innovation research since the data would be far more widely available. This is not considered a substitute but rather a complement to greater accessibility of confidential micro-data to the research community.

- *Expand non-government researcher access to confidential micro-data, including that on business dynamics, while maintaining high standards for confidentiality.*

The Committee recommends that the statistical agencies expand research access to confidential micro-data while maintaining high standards for preserving confidentiality. Outside researcher access to business data, particularly business micro-data, collected by the Federal agencies remains extremely limited. Many scholars are interested in — and with proper confidentiality training and safeguards could be engaged in — research that would further the understanding of different aspects of innovation. While concerns about confidentiality are important and should be at the fore of any discussion on researcher access, current limitations on access are impeding understanding of the dynamic aspects of the economy. Federal agencies do not have adequate funding for in-house staff to exploit fully the research potential of the data. In addition, advances in remote access technologies and the success to date of specific federal agencies in offering outside researcher access make this an opportune time to expand outside researcher access to business micro-data. Academic researchers, subject to confidentiality provisions, could facilitate linkages between confidential micro-data and other existing data such as data collected by the Securities and Exchange Commission, thus providing a richer picture of innovation in the economy. Some agencies have made significant progress in demonstrating the feasibility of expanded access while maintaining strict confidentiality.

4. The Government Should Convene One or More Workshops or Forums under the Auspices of the Secretary of Commerce to Discuss Innovation Drivers, Impediments and Enablers

Many of the members suggested that the development of more effective innovation policies requires undertaking more work to get better identification and examination of innovation drivers, impediments and enablers. Insofar as the Advisory Committee's charge was limited to innovation measurement, these issues have not been directly addressed by the Committee. Nonetheless, given the importance of the issue in the view of most of the Committee members, the Committee is recommending that the Secretary of Commerce convene one or more workshops or forums devoted to examining innovation drivers, impediments and enablers.

Members of the Committee and public commenters identified a broad variety of drivers and enablers of innovation. These included tax policy (R&D tax credits); education policy (promoting science and math); the availability of venture capital; the business climate; ease of entry into and exit from markets; education and skills; research and development expenditures; and specific firm culture and environment. However, in many instances, the correlation between the identified driver or enabler and innovation outcomes is not well understood or convincingly demonstrated empirically.

Impediments to innovation that have been suggested include some immigration policies, trade restrictions, Sarbanes-Oxley rules (particularly Section 404), and other government regulations. As with drivers and enablers, there are many assertions made but the actual impact of such policies on innovation activity is much less clear.

5. The Government Should Continue Participation in the International Dialogue Related to Measuring and Analyzing Innovation and Ensure that U.S. Efforts are Internationally Compatible to the Extent Possible

The Advisory Committee is aware of the work that has been and is being done on innovation measurement in the international community. The Advisory Committee supports the international dialogue on innovation and, indeed, one of the principles adopted by the Committee calls for innovation measures to allow for analysis at international levels.

The United States is actively engaged in international dialogue on innovation and innovation measurement. The National Science Foundation works closely with the Organisation for Economic Co-operation and Development's Working Party of National Experts on Science and Technology Indicators (NESTI). NESTI's mission — to "ensure the continued improvement of the methodologies for the collection of internationally comparable data for measuring the input, output, diffusion and impact of science, technology and innovation (including linkages to economic growth)" — is similar to the mission of this Advisory Committee. The Financial Accounting Standards Board (FASB) and the International Accounting Standards Board (IASB) are working together to converge two sets of standards, which currently differ on the circumstances under which research and development costs can be expensed versus capitalized as intangible assets. The Committee supports such efforts.

6. The Government Should Consider Development of a National Innovation Index when More Work Has Been Done on Both Data Collection and Analysis of Innovation Drivers

An index of innovation or innovation scorecard or dashboard could give some sense of how the U.S. fares on a combination of different measures of innovation. However, even among the members who supported the idea of constructing an index, there was uncertainty as to what measures should be aggregated and how the components should be weighted, if done as a scorecard, and the purpose to which the index would be put. Additionally, while some innovations have immediate and significant economic impact, many innovations appear to have a longer lag before affecting output. As such, a careful consideration of how to weight the long-term and short-term impacts of innovations is necessary prior to implementing an index.

Consideration would need to be given as to the types of measures that would be reported and whether or not data currently are collected on these measures; whether this should be all public data or a mix of public and private information; and whether or not it should be done immediately or be an end product following the development of new innovation measures. Such an index could be a useful tool for international comparisons if there were comparability with measures used in other countries. The development of an innovation index could be useful in the future when more data are available and more is known about specific innovation drivers.

7. The Government Should Support Funding Necessary to Implement the Above Recommendations

Many of the above recommendations will require new funding. The Advisory Committee recommends that the government support the additional funding necessary.

4. How the Business Community Can Help

Innovation is a collaborative process and there is much that the business community — individual firms and business organizations — can do to both assist and drive improvements in innovation measurement. One of the guiding principles endorsed by the Advisory Committee was that innovation data collection efforts should build on the way firms assess the effectiveness of their innovative activities. Insights from individual firms and other organizations can be integrated with those from government efforts to assess innovation activities and performance more comprehensively.

1. The Advisory Committee Recommends that the Business Community Create, Expand and Assess Firm and Industry-Level Measures of Innovation and Develop Best Practices for Innovation Measurement

To provide information on firm and industry-level innovation measures and assess their effectiveness, the Advisory Committee calls on the business community to:

- *Institute firm-level measures of innovation to test the correlation of particular measures with known innovation and to measure innovation in the firm and its impact on firm performance. Possible measures might be based on market share or on innovation intensity.*

Two possible market share measures of innovation were discussed by the Committee. The rationale for looking at a measures based on market share is the assumption that products and services gain market share because they provide better value by being lower priced and/or better satisfying customers' needs. One variation focuses specifically on the share of a firm's new product sales in markets in which the firm is gaining market share. The second variation would measure the extent to which a firm's overall share is growing in a growing market. Individual firms might wish to develop their own market share measures to measure their own innovation performance. In addition, trade associations or other organizations might develop such measures for particular industries and aggregate the results into industry-level measures.

A major issue in designing survey questions based on market share is how to define 'market' in a way that is sufficiently clear to survey respondents to yield a meaningful innovation measure, since a growing market share even in a growing market is not necessarily the result of innovative activity.

Measures of innovation intensity (i.e., the share of firm revenue attributable to recently introduced products and services) may be another useful way to track the impact of innovation. Again, individual firms could track such a measure for their own use or trade associations or other organizations could develop surveys to collect data on total sales revenues and those attributable to products and services introduced to the market in the preceding three years. The data would be used to calculate innovation intensities for each firm in the sample and perhaps could be aggregated to arrive at publishable, industry- and economy-level measures. Researchers could also analyze the resulting firm-level data to better understand the characteristics and performance of firms with different innovation intensities.

Firms with high innovation intensities would presumably be among the most successful in designing, inventing, developing, and implementing new or altered products and services. However, there are formidable interpretive challenges. A 'new' product or service may be 'new' only to the firm and it could be difficult to differentiate truly innovative activity. 'New to market' in manufacturing may only capture product innovations; while 'new to firm' in service industries may be an indirect indicator of innovations in processes, systems, organizational structures, or business models. At least initially, such measures may be most useful for individual firms to develop and test the usefulness of their own innovation intensities.

The Organization for Economic Cooperation and Development (OECD) has designed survey questions to collect data for the development of measures of innovation intensity. Several countries have implemented such surveys (e.g., the EU Community Innovation Survey), and the U.S. business community could build on that experience.

The Committee calls on the business community, particularly individual firms and trade associations, in cooperation with the research community, to experiment with collecting data on market share- based and innovation intensity measures, analyze the relationship between such measures and firm performance and report on the findings. Insights from such analyses could be integrated with those from government and academia to provide a comprehensive view of innovation in the economy.

- *Develop and implement best practices in innovation management and accounting.*

The development of best practices in innovation management and accounting appears to be in its infancy. The Committee calls on the business community, particularly trade associations, to collect information about current innovation management and accounting practices, and calls on universities to develop curricula and research programs in innovation management and accounting. The government can help academic research programs in this area by supporting such efforts through agencies such as the National Science Foundation.

2. The Advisory Committee Recommends that the Business Community Participate In Research Activities and, as Appropriate, Make Innovation Information Available to Researchers

Sharing knowledge about innovation and participating in the development of it is a way the business community can help drive the way toward better innovation measurement. In particular, the Advisory Committee recommends that firms and associations:

- *Participate in collaborative projects as a means of assembling a broad range of data related to innovation.*

Private sector initiatives to collect and analyze data on innovation activities and outcomes can complement and help inform government data collection and publication efforts as well as help identify and disseminate best practices in innovation management.

- *File public reports in XBRL format when it is an option.*

As noted elsewhere in the report, firms could facilitate measurement and analysis of innovation activities and outcomes by reporting financial and other information in XBRL format. The Committee calls on firms to report information in XBRL format whenever possible.

5. Where Research is Needed

Much more needs to be learned about measuring innovation and its impact. Knowledge about innovation would be greatly enhanced if government, academic and private sector researchers— alone and in collaborative efforts — were to undertake research in the following areas:

1. Identification and Assessment of Innovation Outcome Measures

The definition for innovation adopted by the Advisory Committee focuses on innovation outcomes. Much of the innovation measurement currently focuses solely on inputs to innovation. More work needs to be done to define appropriate outcome measures and analyze their utility and effectiveness. Particular research areas include:

- *Assessment of the effectiveness of measures based on market share as innovation measures and the feasibility, cost, and burden of developing such measures.*

The Advisory Committee has called on the business community to develop and test measures based on market share. Researchers, too, need to examine these measures. The Committee calls on researchers to assess the feasibility, cost, and effectiveness of collecting data to develop measures based on market share, and, where appropriate, to sponsor or participate in pilot projects to test such measures.

- *Assessment of the feasibility, cost, and burden of developing measures of innovation intensity, including a review of other countries' experience in this area and consideration of a pilot project.*

High innovation intensity (the share of revenues earned by recently introduced products and services) is also considered to be a possible indicator of successful innovation. The Committee calls for research to assess the feasibility, cost, and effectiveness of collecting data to estimate innovation intensity measures, and, where appropriate, to sponsor or participate in pilot projects to test such measures.

- *Analysis of the qualitative and quantitative impacts of specific drivers, impediments and enablers on innovation outcomes.*

Work needs to be done to determine the actual impacts of particular drivers, enablers and impediments — both internal to firms (such as firm culture) and externally imposed (such as accounting rules and immigration policies). In many instances, the connection between an identified driver, enabler or impediment and innovation outcomes is neither well understood nor convincingly demonstrated empirically.

2. Identification of GAPs in Innovation Data and How they Might Be Filled

There are many gaps in the collection of data that might explain innovation better. The Advisory Committee urges work in the following areas:

- *Identification of new data that would be useful in measuring innovation.*

Simply implementing the Committee's proposals for data collection is unlikely to answer all questions about innovation. The process of implementing the proposals will raise new questions and stimulate new ideas for measuring various facets of innovation. The Committee urges that researchers continually probe the data generated by this project to discover ways to improve the data or propose new kinds of data that would help fulfill the Committee's mission, and identify data the Committee has proposed collecting that should be abandoned as not cost-effective with respect to the Committee's mission.

- *Assessment of the feasibility, cost and burden of collecting data on intellectual property transactions.*

The Committee has proposed the collection of more complete data on firms' total expenditures on and revenues from intellectual property transactions (including both domestic as well as cross-border transactions). In addition, the Committee urges researchers to assess the feasibility, cost and burden of collecting such data not just at the firm level but also at the transaction level by contracting with a representative sample of buyers and sellers in the academic, non-profit, and private sectors to compile periodic data on transactions involving technology licensing. It may be appropriate to carry out a pilot study to assess the nature and extent of the data available.

Measuring this "innovations market" is vital for measuring innovation in a knowledge economy. Insofar as innovation results in the creation of intangible knowledge assets, which yield a flow of benefits over time, measuring transactions in intangible knowledge assets is a good way to measure the value of these assets. Further, transactions data would complement the proposed new NIPA architecture by providing more accurate measures of purchases of the services of intangible inputs. Given that such purchases have grown very substantially in the last two decades, ignoring them risks the very objective that the new NIPA architecture is to accomplish. Finally, transactions data would complement the proposal to improve measures of services industries; many services firms play a crucial role in generating and diffusing innovations through the economy.

- *Identification of ways to overcome gaps and shortcomings in historical measures of intangible investments.*

The proposals on measuring intangible assets discussed elsewhere in this chapter focus mainly on collecting data on new investments in intangibles or estimates of the market value of intangible capital stocks. Absent good data on the market value of intangible capital stocks, however, it may be necessary to develop historical cost estimates in order to account for the contribution of intangible capital stocks to productivity growth. The Committee encourages researchers to develop methods for estimating past investments in intangibles.

3. Analysis of Relationships between Innovation Activities and Collaboration, Innovation Performance, and Firm Performance

A major goal of innovation measurement is to develop estimates of the rate of return to investments in innovation activities.

A methodological challenge is to untangle the direction of causality between activities and innovation performance. Suppose researchers find that increasing innovation activity X (e.g., spending on a broad set of innovation activities, including but not limited to R&D) is associated with an increase in firm innovation performance Y (e.g., number of new trademarks). And suppose researchers also found that increasing firm innovation performance Y is associated with an increase in firm financial performance measure Z (e.g., sales revenue). Does this mean that X causes Z? Or do growing firms simply tend to spend more on innovation activities?

Another methodological and data challenge is dealing with the fact that much innovation is not done strictly within a firm's boundaries — firms collaborate with other firms and with universities; firms purchase and sell intellectual property rights; firms outsource some innovation activities — so getting the whole picture of innovation is key to estimating the rate of return to innovation activities. And, indeed, innovative activity happens in an international setting where many of these interactions cross national borders.

- *Analysis of the relationship between innovation and occupational employment using firm-level micro-data.*

Some observers have suggested that the United States faces shortages of workers in scientific, engineering, and technical (SET) occupations that will harm our ability to compete with other countries in the global marketplace. Others have suggested that firms in the United States can leverage their SET employees by outsourcing some scientific R&D and other innovation-related work or by increasing immigration of SET workers. Still others have suggested that competitiveness depends on a broader range of occupations that includes other kinds of creative workers (e.g., artists and designers) making it harder to say definitively that the United States is `falling behind' other economies when it comes to the supply of innovation workers.

A possible research area using the data generated by implementing the Committee's proposals is to explore the relationship between occupational employment — of SET workers and more broadly defined 'innovation workers' — and various measures of innovation outputs such as intellectual property (patents, copyrights, trademarks), intellectual property licensing revenue and expenditure flows, innovation intensity, and measures of firm performance.

- *Evaluation of whether firms with high innovation intensities perform better than otherwise similar firms with low intensities.*

This is perhaps the most direct measure of how the introduction of new products and services is associated with firm performance. It provides a key link that helps explain how investments in innovation may yield better firm performance. It does so by bridging the gap between investment and financial results to get at how the introduction of new products has an impact on firm profitability.

- *Assessment of the effect of collaboration on innovation outcomes and identification of the key elements of successful collaborative activities.*

Collaboration is believed to be a key element in fostering innovation yet little research has been done to confirm this belief. Work needs to be done both to develop methods to analyze collaboration and assess its actual impact.

- *Assessment and analysis of cross-national innovation activities of firms.*

In today's globally connected world, many U.S. multinationals conduct a significant part of their innovation activities outside the United States. Conversely, many non-U.S. firms conduct a significant part of their innovation activities in the United States. These cross-national innovation activities undoubtedly affect the U.S. economy. The statistical agencies are currently working to link data on domestic and international research and development expenses by U.S. and foreign multinational firms. Such efforts should be encouraged and expanded to permit examination of the output of cross-national innovation activities.

- *Analysis of publicly filed financial and other data on firms, particularly as the data become more user-friendly, to identify innovative practices and firms.*

Publicly filed financial data could be used to examine characteristics of firms considered innovative and to consider whether these innovative firms have unique practices or characteristics.

- *Description and explanation of business dynamics and to the extent possible, analysis of their relationship to innovation using longitudinal business databases.*

Studying the entry, growth, and exit of establishments and firms could provide important information on the characteristics of successful and unsuccessful firms. Much innovation occurs in the smallest firms, where high exit and entry rates are common. Evidence for several industries suggests that entering firms are, on average, more productive than exiting firms but the role played by innovation is not known.

The Census Bureau is developing a longitudinal business database from which it is planning to publish data on establishment and firm births and deaths, job creation and destruction by firm size, age, and industrial sector, and other measures of business dynamics over the past three decades. The resulting data will help researchers distinguish between the effects of short-term economic fluctuations and long-term effects (such as innovation) on business dynamics.

Analysis of longitudinal business dynamics data is believed to be able to provide insight into the patterns of innovative activity in the economy and the innovative and entrepreneurial environment in various time periods. New data sets such as the Kauffman Firm Survey show potential for understanding how high-impact firms evolve, but longer business panel surveys and additional government support are needed to exploit their full potential.

The Committee encourages the Census Bureau to continue its efforts to publish reports on business dynamics and explore the possibility of developing public use micro-data on business dynamics. The Committee also encourages researchers to use longitudinal business databases to help describe and explain business dynamics and, to the extent possible, analyze its relationship to innovation.

- *Exploration of the use of different sources of available data to determine whether there are correlations between innovative performance of firms and the existence of a new regulator), environment.*

Regulatory policies are believed to impact innovation — both positively and negatively — but there is little or no empirical evidence of their impact.

Conclusion

The Advisory Committee's recommendations, if adopted, will go far in setting this nation on a course toward effectively measuring the impact of innovation on the economy. The work is essential and the Committee calls on the government, the business community and researchers to adopt the recommendations.

End Notes

[1] Total factor productivity (TFP) is output per unit of total input (hours worked and use of capital). The growth of TFP in excess of the growth of total inputs is attributable, in part, to innovation.

[2] Schramm, C. J. 2006. The Entrepreneurial Imperative. Harper Collins.

[3] Baumol, W. R Litan, and C. Schramm. 2007. *Good Capitalism, Bad Capitalism and the Economics of Growth and Prosperity:* Yale, 2007.

[4] The seminal work on this subject is found in Robert Solow, "A Contribution to the Theory of Economic Growth." *Quarterly Journal of Economics* 70:65-94 (1956); and Edward F. Denison, "Sources of Economic Growth in the United States and the Alternatives before Us." (New York: Committee for Economic Development, 1962). For a more recent treatment, see Baumol, Landes, & Mokyr, *Entrepreneurship and Economic History.* Princeton (forthcoming).

[5] Thomas, David G. 2006. *Blueprint to a Billion: 7 Essentials to Achieve Exponential Growth.* John Wiley & Sons

CHAPTER SOURCES

The following chapters have been previously published:

Chapter 1 – This is an edited, excerpted and augmented edition of a United States Congressional Research Service publication, Report Order Code R40677, dated June 29, 2009.

Chapter 2 – This is an edited, excerpted and augmented edition of a United States Congressional Research Service publication, Report Order Code RS22815, dated June 14, 2010.

Chapter 3 – This is an edited, excerpted and augmented edition of a United States Congressional Research Service publication, Report Order Code 96-402, dated August 4, 2010.

Chapter 4 – This is an edited, excerpted and augmented edition of a United States Congressional Research Service publication, Report Order Code RS22865, dated August 4, 2010.

Chapter 5 – This is an edited, excerpted and augmented edition of a United States Congressional Research Service publication, Report Order Code RL34559, dated January 6, 2010.

Chapter 6 – This is an edited, excerpted and augmented edition of a report to the Secretary of Commerce by the Advisory Committee on Measuring Innovation in the 21st Century Economy, dated January 2008.

INDEX

D

E

F

G

H

I

J

K

L

M

N

O

P

Q

R

S

T

U

V

W

Y